Textbook of Polymer Chemistry

Textbook of Polymer Chemistry

Brent Burton

Willford Press
www.willfordpress.com

Published by Willford Press,
118-35 Queens Blvd., Suite 400,
Forest Hills, NY 11375, USA

ISBN: 978-1-64728-015-4

Cataloging-in-Publication Data

Textbook of polymer chemistry / Brent Burton.
p. cm.
Includes bibliographical references and index.
ISBN 978-1-64728-015-4
1. Polymers. 2. Polymerization. 3. Chemistry. I. Burton, Brent.
QD381 .T49 2022
547.7--dc23

For information on all Willford Press publications
visit our website at www.willfordpress.com

Willford Press

Contents

Preface

A polymer is a large macromolecule, which is formed by the process of polymerization of monomers. The sub-discipline of chemistry that focuses on the chemical synthesis, physical and chemical properties, and the structure of macromolecules and polymers is referred to as polymer chemistry. It primarily deals with synthetic and organic compositions. Synthetic polymers are used in commercial materials and products such as plastics and rubbers. They are also major components of composite materials. Some of the techniques used within this field are free-radical polymerization, ionic polymerization, polycondensation, protein biosynthesis and cell-free protein synthesis. This textbook provides comprehensive insights into the field of polymer chemistry. Different approaches, evaluations, methodologies and techniques related to this field have been included in this book. It is appropriate for those seeking detailed information in this area.

A short introduction to every chapter is written below to provide an overview of the content of the book:

Chapter 1 - The chained molecules that are composed of many repeated subunits are known as polymers. Polymer chemistry is the sub-discipline of chemistry which is concerned with the structure, chemical synthesis, physical and chemical properties of polymers. This is an introductory chapter which will briefly introduce all the significant aspects of polymers and polymer chemistry; **Chapter 2** - Polymers are classified on the basis of source, structure, mode of polymerization, molecular forces, etc. It includes rubber polymers, biological polymers, synthetic polymers, linear polymers, addition polymers, elastomers, etc. This chapter has been carefully written to provide an easy understanding of these different types of polymers; **Chapter 3** - Organic polymers contain carbon atoms and inorganic polymers have ionic and hydrocarbon bonds but do not contain carbon atoms. Olysiloxanes, polyphosphazenes, polysilanes, etc. are some inorganic polymers and polytetrafluoroethylene, fluoropolymers, etc. are a few organic polymers. This chapter discusses about these organic and inorganic polymers in detail; **Chapter 4** - Polymerization is the reaction between molecules in a chemical reaction in order to form three-dimensional network or polymer chains. The major types of polymerization are coordination polymerization, dispersion polymerization, step-growth polymerization, nitroxide-mediated radical polymerization, ring-opening polymerization, etc. All these types of polymerization have been carefully analyzed in this chapter; **Chapter 5** - Stereochemistry refers to the discipline of chemistry that studies the relative spatial arrangement of atoms which form the structure of molecules. This chapter closely examines the key concepts of stereochemistry of polymers to provide an extensive understanding of the subject.

Finally, I would like to thank my fellow scholars who gave constructive feedback and my family members who supported me at every step.

Brent Burton

1

Understanding Polymers and Polymer Chemistry

The chained molecules that are composed of many repeated subunits are known as polymers. Polymer chemistry is the sub-discipline of chemistry which is concerned with the structure, chemical synthesis, physical and chemical properties of polymers. This is an introductory chapter which will briefly introduce all the significant aspects of polymers and polymer chemistry.

Polymer

Polymers are materials made of long, repeating chains of molecules. The materials have unique properties, depending on the type of molecules being bonded and how they are bonded. Some polymers bend and stretch, like rubber and polyester. Others are hard and tough, like epoxies and glass.

Polymers touch almost every aspect of modern life. Chances are most people have been in contact with at least one polymer-containing product — from water bottles to gadgets to tires — in the last five minutes. The term polymer is often used to describe plastics.

Characteristics of Polymers

Every polymer has very distinct characteristics, but most polymers have the following general attributes:

- Polymers can be very resistant to chemicals: Consider all the cleaning fluids in your home that are packaged in plastic. Reading the warning labels that describe what happens when the chemical comes in contact with skin or eyes or is ingested will emphasize the need for chemical resistance in the plastic packaging. While solvents easily dissolve some plastics, other plastics provide safe, non-breakable packages for aggressive solvents.
- Polymers can be both thermal and electrical insulators: A walk through your house will reinforce this concept, as you consider all the appliances, cords, electrical outlets and wiring that are made or covered with polymeric materials. Thermal resistance is evident in the kitchen with pot and pan handles made of polymers, the coffee pot handles, the foam core of refrigerators and freezers, insulated cups, coolers, and microwave cookware. The thermal underwear that many skiers wear is made of polypropylene and the fiberfill in winter jackets is acrylic and polyester.
- Generally, polymers are very light in weight with significant degrees of strength: Consider the range of applications, from toys to the frame structure of space stations, or from

delicate nylon fiber in pantyhose to Kevlar, which is used in bulletproof vests. Some polymers float in water while others sink. But, compared to the density of stone, concrete, steel, copper, or aluminum, all plastics are lightweight materials.

- Polymers can be processed in various ways: Extrusion produces thin fibers or heavy pipes or films or food bottles. Injection molding can produce very intricate parts or large car body panels. Plastics can be molded into drums or be mixed with solvents to become adhesives or paints. Elastomers and some plastics stretch and are very flexible. Some plastics are stretched in processing to hold their shape, such as soft drink bottles. Other polymers can be foamed like polystyrene (Styrofoam), polyurethane and polyethylene.
- Polymers are materials with a seemingly limitless range of characteristics and colors: Polymers have many inherent properties that can be further enhanced by a wide range of additives to broaden their uses and applications. Polymers can be made to mimic cotton, silk, and wool fibers; porcelain and marble; and aluminum and zinc. Polymers can also make possible products that do not readily come from the natural world, such as clear sheets and flexible films.
- Polymers are usually made of petroleum, but not always: Many polymers are made of repeat units derived from natural gas or coal or crude oil. But building block repeat units can sometimes be made from renewable materials such as polylactic acid from corn or cellulosics from cotton linters. Some plastics have always been made from renewable materials such as cellulose acetate used for screwdriver handles and gift ribbon. When the building blocks can be made more economically from renewable materials than from fossil fuels, either old plastics find new raw materials or new plastics are introduced.
- Polymers can be used to make items that have no alternatives from other materials: Polymers can be made into clear, waterproof films. PVC is used to make medical tubing and blood bags that extend the shelf life of blood and blood products. PVC safely delivers flammable oxygen in non-burning flexible tubing. And anti-thrombogenic material, such as heparin, can be incorporated into flexible PVC catheters for open heart surgery, dialysis, and blood collection. Many medical devices rely on polymers to permit effective functioning.

Properties of Polymers

Physical Properties

Physical properties of polymers include molecular weight, molar volume, density, degree of polymerization, crystallinity of material, and so on.

Degree of Polymerization and Molecular Weight

The degree of polymerization (DP)-n in a polymer molecule is defined as the number of repeating units in the polymer chain. For example,

$$-(-CH_2 - CH_2-)-_n$$

The molecular weight of a polymer molecule is the product of the degree of polymerization and the molecular weight of the repeating unit. The polymer molecules are not identical but are a mixture of many species with different degrees of polymerization, that is, with different molecular weights. Therefore, in the case of polymers we talk about the average values of molecular weights.

Molecular Weight Averages: Suppose we have a set of values $\{x_1, x_2, ..., x_n\}$ and the corresponding probability of occurrence is given by $\{P_1, P_2, ..., P_n\}$, then the average value is defined as follows:

$$\sum_{i=0}^{\infty} P_i x_i$$

Number-Average Molecular Weight: If N_i is the number of polymer molecules having the molecular weight M_i, then the "number-average" probability of the given mass is given by:

$$P_i = \frac{N_i}{\sum_{j=0}^{\infty} N_j}$$

The number-average molecular weight is given by:

$$M_n = \sum_{i=0}^{\infty}\left[\frac{N_i}{\sum_{j=0}^{\infty} N_j}\right] M_i = \frac{\sum_{i=0}^{\infty} M_i N_j}{\sum_{j=0}^{\infty} N_j}$$

The physical properties (such as transition temperature, viscosity, etc.) and mechanical properties (such as strength, stiffness, and toughness) depend on the molecular weight of polymer. The lower the molecular weight, lower the transition temperature, viscosity, and the mechanical properties. Due to increased entanglement of chains with increased molecular weight, the polymer gets higher viscosity in molten state, which makes the processing of polymer difficult.

Weight-Average Molecular Weight: The weight-average probability is given by:

$$P_i = \frac{N_i M_i}{\sum_{j=0}^{\infty} N_j M_j}$$

The weight-average molecular weight is given by:

$$M_w = \sum_{i=0}^{\infty}\left[\frac{N_i M_i}{\sum_{j=0}^{\infty} N_j M_j}\right] M_i = \frac{\sum_{i=0}^{\infty} N_i M_i^2}{\sum_{j=0}^{\infty} N_j M_j}$$

A typical plot showing the number-average and weight-average molecular weight is shown in figure. The number-average molecular weight is less than the weight-average molecular weight. The degree of polymerization can be calculated using the number-average molecular weight.

$$\text{Degree of polymerization} = \frac{\text{Number average molecular weight}}{\text{Molecular weight of the repeat unit}}$$

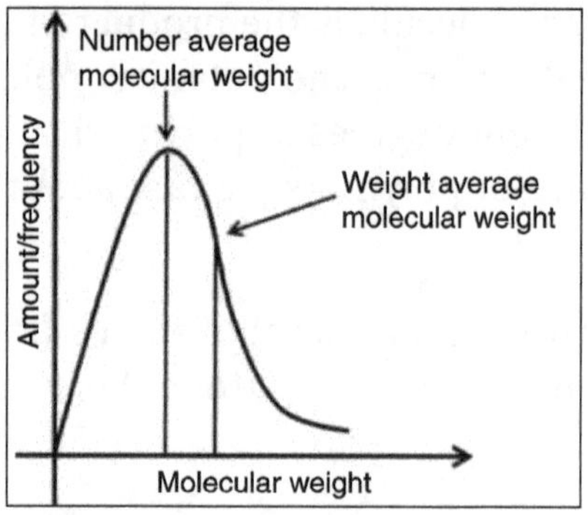

Average molecular weights of polymer.

Polydispersity Index or Heterogeneity Index: The ratio of the weight-average molecular weights to the number-average molecular weights is called polydispersity index (PDI) or heterogeneity index, which measures the polydispersity of the polymer mixture.

$$\text{PDI} = \frac{M_w}{M_n}$$

The dispersity measures heterogeneity of sizes of molecules or particles in the mixture. The mixture is called monodisperse if the molecules have the same size, shape, or mass. If the molecules in the mixture have an inconsistent size, shape and mass distribution, the mixture is called polydisperse.

The natural polymers are generally monodisperse as all synthetic polymers are polydisperse with some exceptions. The PDI is equal to or greater than 1 whereas the polymer chains approach uniform chain length, the PDI tends to unity.

Polymer Crystallinity: Crystalline and Amorphous Polymers

The polymeric chains being very large are found in the polymer in two forms as follows:

Lamellar crystalline form in which the chains fold and make lamellar structure arranged in the regular manner and amorphous form in which the chains are in the irregular manner. The lamellae are embedded in the amorphous part and can communicate with other lamellae via tie molecules. Polymer may be amorphous or semi-crystalline in nature.

The %crystallinity is given by:

$$\%\text{Crystallinity} = \frac{\rho_c(\rho_s - \rho_a)}{\rho_s(\rho_c - \rho_a)} \times 100$$

ρ_c =density of the completely crystalline polymer,

ρ_a =density of the completely amorphous polymer,

ρ_s =density of the sample.

A typical range of crystallinity can be defined as amorphous (0%) to highly crystalline (>90%). The polymers having simple structural chains as linear chains and slow cooling rate will result in good crystallinity as expected. In slow cooling, sufficient time is available for crystallization to take place. Polymers having high degree of crystallinity are rigid and have high melting point, but their impact resistance is low. However, amorphous polymers are soft and have lower melting points. For a solvent, it is important to state that it can penetrate the amorphous part more easily than the crystalline part.

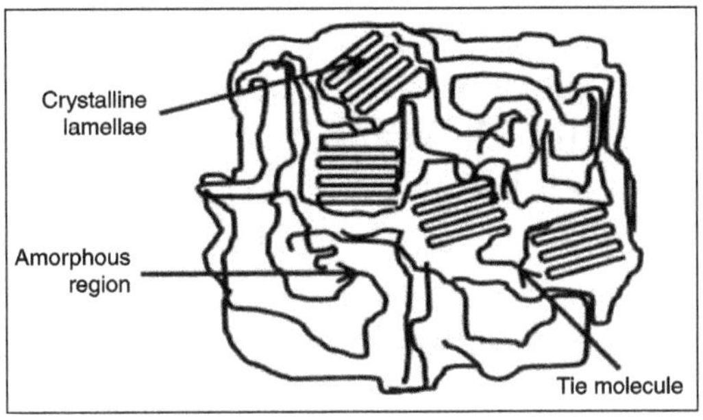

Semi-crystalline polymer.

Examples of amorphous polymers: polystyrene and poly (methyl methacrylate). Examples of crystalline polymers: polyethylene, and PET polyester.

Spherulites: if the molten polymer is cooled down, then the crystalline lamellae grow in radial direction from a nucleus along the three dimensions leading to a spherical structure called spherulite. The amorphous region is in between the crystalline lamellae. Spherulite formation and its diameter depend on various parameters such as the number of nucleation sites, polymer molecule structure and rate of cooling. Due to highly ordered lamellae in the spherulite, it shows higher density, hardness, tensile strength, and Young's modulus. The elasticity and impact resistance are shown, because the lamellae are connected to amorphous regions.

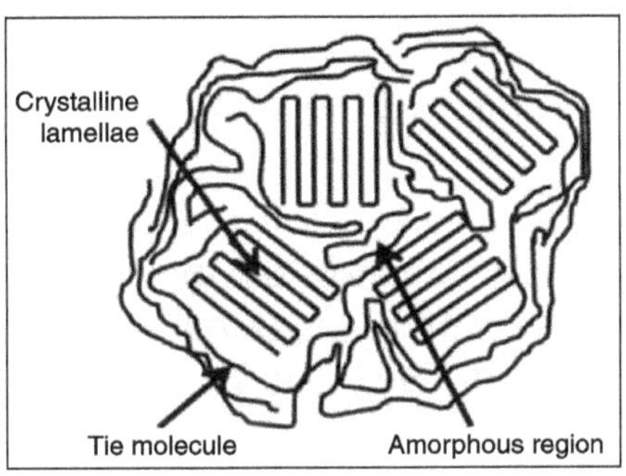

A typical structure of spherulite.

Thermal Properties of Polymers

In the amorphous region of the polymer, at lower temperature, the molecules of the polymer are in, say, frozen state, where the molecules can vibrate slightly but are not able to move significantly. This state is referred as the glassy state. In this state, the polymer is brittle, hard and rigid analogous to glass. Hence the name glassy state. The glassy state is similar to a supercooled liquid where the molecular motion is in the frozen state. The glassy state shows hard, rigid, and brittle nature analogous to a crystalline solid with molecular disorder as a liquid. Now, when the polymer is heated, the polymer chains are able to wiggle around each other, and the polymer becomes soft and flexible similar to rubber. This state is called the rubbery state. The temperature at which the glassy state makes a transition to rubbery state is called the glass transition temperature T_g. Note that the glass transition occurs only in the amorphous region, and the crystalline region remains unaffected during the glass transition in the semi-crystalline polymer.

Melting Point and Glass Transition Temperature

The glass transition temperature is the property of the amorphous region of the polymer, whereas the crystalline region is characterized by the melting point. In thermodynamics, the transitions are described as first and second order transitions. Glass transition temperature is the second order transition, whereas the melting point is the first order transition. The value of glass transition temperature is not unique because the glassy state is not in equilibrium. The value of glass transition temperature depends on several factors such as molecular weight, measurement method, and the rate of heating or cooling. Approximate values of glass transition temperatures of some polymers are listed in table.

The semi-crystalline polymer shows both the transitions corresponding to their crystalline and amorphous regions. Thus, the semi-crystalline polymers have true melting temperatures (T_m) at which the ordered phase turns to disordered phase, whereas the amorphous regions soften over a temperature range known as the glass transition (T_g). It should be noted that amorphous polymers do not possess the melting point, but all polymers possess the glass transition temperature.

The polymer melting point T_m is increased if the double bonds, aromatic groups, bulky or large side groups are present in the polymer chain, because they restrict the flexibility of the chain. The branching of chains causes the reduction of melting point, as defects are produced because of the branching.

Factors affecting the Glass Transition Temperature

The glass transition temperature depends on the mobility and flexibility (ease of the chain segment to rotate along the chain backbone) of the polymeric chains. If the polymeric chains can move easily, then the glassy state can be converted to the rubbery state at lower temperature, that is, the glass transition temperature is lower. If somehow the mobility of the chains is restricted, then the glassy state is more stable, and it is difficult to break the restriction causing the immobility of the polymer chains at the lower temperature, because more energy is required to make the chains free. Thus, in this case, the glass transition temperature is raised.

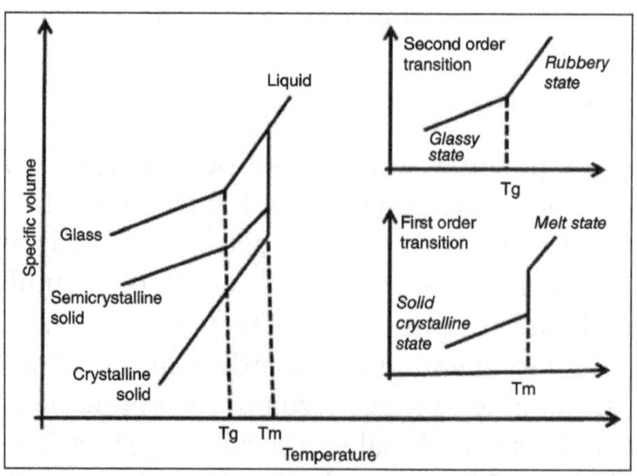

Melting point and glass transition temperature of polymer.

Table: Glass Transition Temperatures of Some Polymers.

Polymer	T_g (°C)
Polytetrafluoroethylene	−97
Polypropylene (isotactic)	+100
Polystyrene	+100
Poly(methylmethacrylate) (atactic)	+105
Nylon 6,6	+57
Polyethylene (LDPE)	−120
Polyethylene (HDPE)	−90
Polypropylene (atactic)	−18
Polycarbonate	+150
Poly(vinyl acetate) (PVAc)	+28
Polyester(PET)	+69
Poly(vinyl alcohol) (PVA)	+85
Poly(vinyl chloride) (PVC)	+87

- Intermolecular Forces: Strong intermolecular forces cause higher T_g. For example, PVC (T_g = 80 °C) has stronger intermolecular forces than polypropylene (T_g = −18 °C) because of the dipole – dipole forces from the C—Cl bond.
- Chain Stiffness: The presence of the stiffening groups (such as amide, sul- fone, carbonyl, p-phenylene etc.) in the polymer chain reduces the flexibility of the chain, leading to higher glass transition temperature. For example, polyethyleneterephthalete is stiffer than polyethylene adipate due to the presence of benzene ring. Therefore, T_g value is higher for polyethyleneterephthalate.
- Cross-Linking: The cross-links between chains restrict rotational motion and raise the glass transition temperature. Hence, higher cross-linked molecule will show higher T_g than that with lower cross-linked molecule.
- Pendant groups: The presence of pendent group can change the glass transition temperature.
    - Bulky pendant groups: The presence of bulky pendant group, such as a benzene ring, can restrict rotational freedom, leading to higher glass transition temper- ature. As in polystyrene, the presence of benzene ring increases the T_g. In polypropylene, there is no benzene ring that leads to lower T_g value.
    - Flexible pendant groups: The presence of flexible pendant groups, for example, aliphatic chains, limits the packing of the chains and hence increases the rotational motion, tending to less Tg value. In polybutylmethacrylate, the presence of large aliphatic chain reduces the Tg value when compared with that of polymethylmethacrylate.

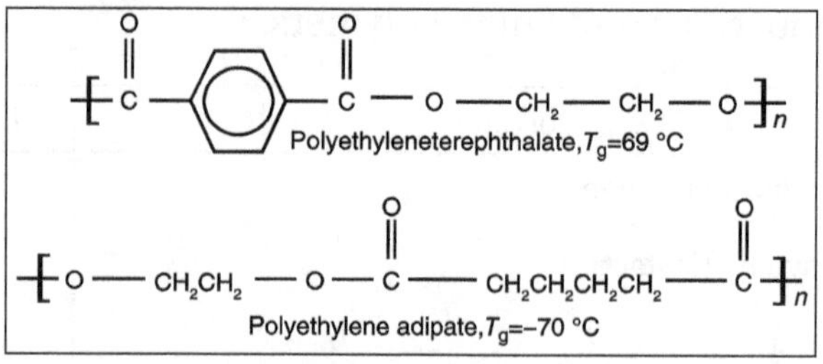

Presence of benzene in polyethyleneterephthalete (top molecular chain) makes it stiffer (hence higher T_g) than polyethylene adipate (bottom molecular chain).

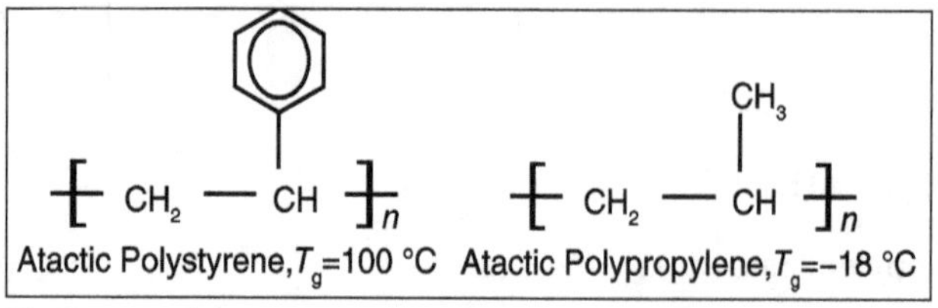

Role of bulky pendant groups in affecting glass transition temperature.

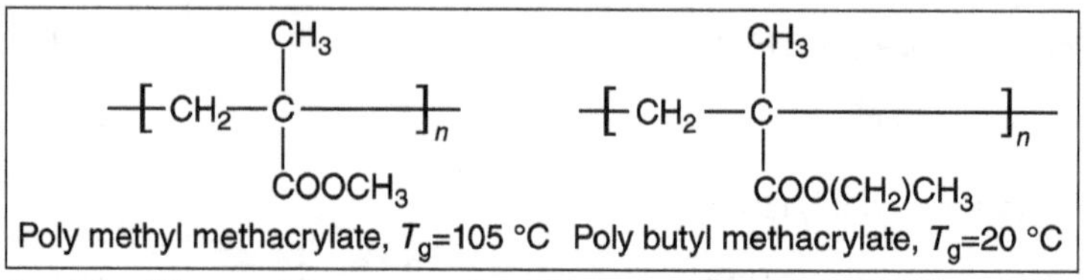

Role of flexible pendant groups in affecting glass transition temperature.

- Plasticizers: Plasticizers are low molecular weight and non-volatile mate- rials added to polymers to increase their chain flexibility. They reduce the intermolecular cohesive forces between the polymer chains, which in turn decrease T_g.
- Molecular Weight: The glass transition temperature is also affected by the molecular weight of the polymer. T_g is increased with the molecular weight. The molecular weight is related to the glass transition temperature by the Fox–Flory Equation:

$$T_g = T_{g,\infty} - \frac{K}{M_n}$$

where $T_{g,\infty}$ is the glass transition temperature at the molecular weight of infinity, and K is the empirical parameter called Fox–Flory parameter related to the free volume inside the polymer. It is observed that T_g is increased up to the molecular weight of approximately 20 000, and after this limit, the T_g is not affected appreciably.

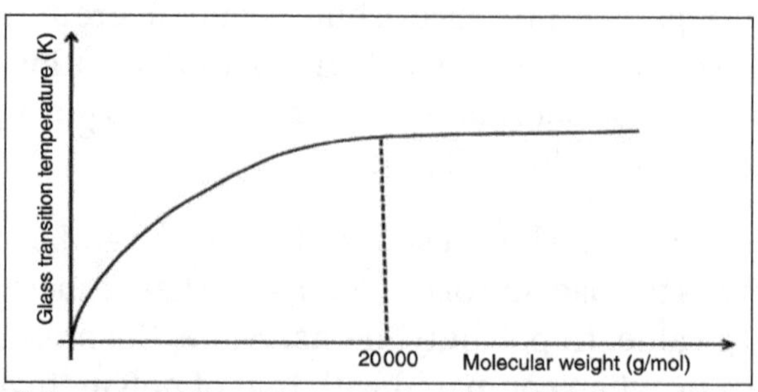

Variation of glass transition temperature with molecular weight of polymer.

Mechanical Properties

It is of great importance to be familiar with some basic mechanical properties of the material before its application in any field, such as how much it can be stretched, how much it can be bent, how hard or soft it is, how it behaves on the application of repeated load and so on.

1. Strength: In simple words, the strength is the stress required to break the sample. There are several types of the strength, namely tensile (stretching of the polymer), compressional (compressing the polymer), flexural (bending of the polymer), torsional (twisting of the polymer), impact (hammering) and so on. The polymers follow the following order of increasing strength: linear < branched < cross-linked < network.

Factors Affecting the Strength of Polymers

- Molecular Weight: The tensile strength of the polymer rises with increase in molecular weight and reaches the saturation level at some value of the molecular weight. The tensile strength is related to molecular weight by the following equation.

$$\sigma = \sigma_{\infty} - \frac{A}{M}$$

σ_{∞} is the tensile strength of the polymer with molecular weight of infinity. A is some constant, and M is the molecular weight. At lower molecular weight, the polymer chains are loosely bonded by weak van der Waals forces and the chains can move easily, responsible for low strength, although crystallinity is present. In case of large molecular weight polymer, the chains become large and hence are entangled, giving strength to the polymer.

- Cross-linking: The cross-linking restricts the motion of the chains and increases the strength of the polymer.

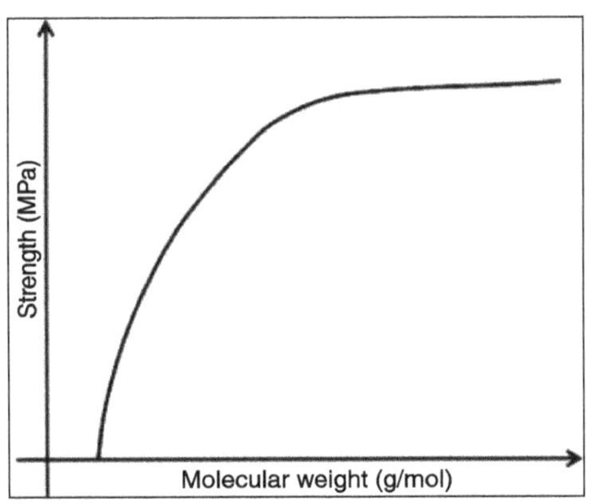

Variation of tensile strength with molecular weight of the polymer.

- Crystallinity: The crystallinity of the polymer increases strength, because in the crystalline phase, the intermolecular bonding is more significant. Hence, the polymer deformation can result in the higher strength leading to oriented chains.

2. Percent Elongation to Break (Ultimate Elongation): It is the strain in the material on its breakage, as shown in figure below. It measures the percentage change in the length of the material

before fracture. It is a measure of ductility. Ceramics have very low (100%), thermosets (5%) value of elongation to break.

3. Young's Modulus (Modulus of Elasticity or Tensile Modulus): Young's Modulus is the ratio of stress to the strain in the linearly elastic region. Elastic modulus is a measure of the stiffness of the material.

$$E = \frac{\text{Tensile Stress}(\sigma)}{\text{Tensile Strain}(\varepsilon)}$$

4. Toughness: The toughness of a material is given by the area under a stress–strain curve.

$$\text{Toughness} = \int \sigma \, d\varepsilon$$

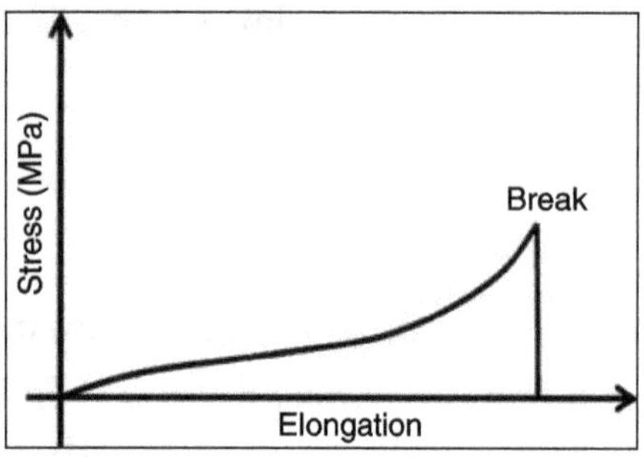

Elongation to break of the polymer.

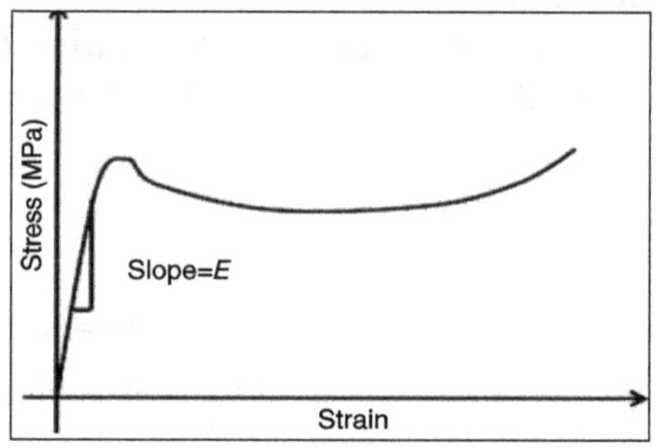

Young's modulus of the polymer.

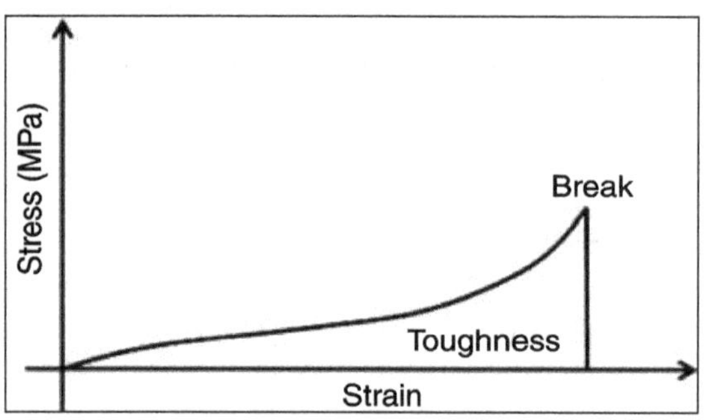

The toughness of polymer material.

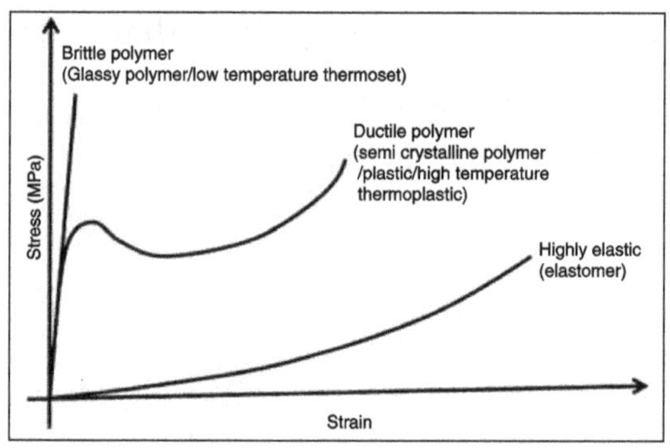

Stress–strain behavior of different types of materials.

The toughness measures the energy absorbed by the material before it breaks.

A typical stress–strain curve is shown in figure above, which compares the stress–strain behavior of different types of materials. The rigid materials possess high Young's modulus (such as brittle polymers), and ductile polymers also possess similar elastic modulus, but with higher fracture toughness. However, elastomers have low values of Young's modulus and are rubbery in nature.

The yield strength of the plastic polymer is the corresponding stress where the elastic region (linear portion of the curve) ends. The tensile strength is the stress corresponding to the fracture of the polymer. The tensile strength may be higher or lower than the yield strength.

The mechanical properties of the polymer are strongly affected by the temperature. A typical plot of stress versus strain is shown in figure below. From the plot, it is clear that with increase in the temperature, the elastic modulus and tensile strength are decreased, but the ductility is enhanced.

5. Viscoelasticity: There are two types of deformations: elastic and viscous. Consider the constant stress level applied to a material as shown in the figure below.

In the elastic deformation, the strain is generated at the moment the constant load (or stress) is applied, and this strain is maintained until the stress is not released.

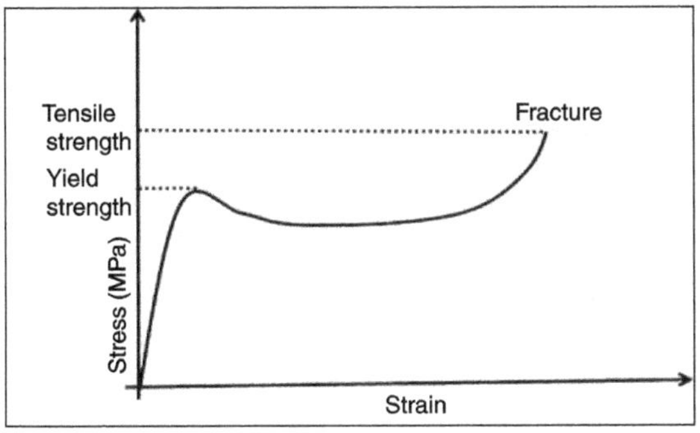

Yield strength and tensile strength of polymer.

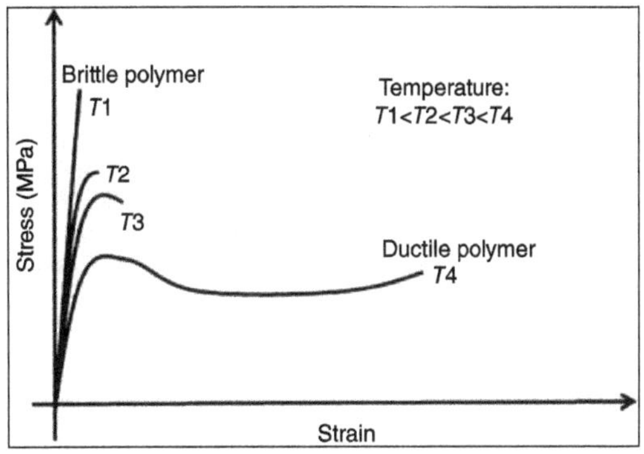

Effect of temperature on the mechanical properties of polymer.

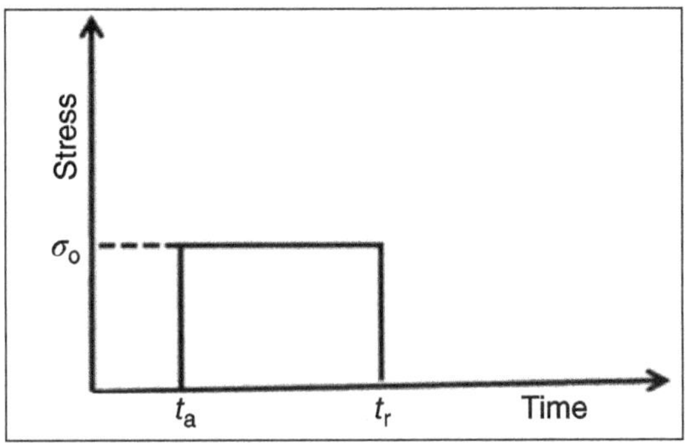

Constant stress applied to a polymer.

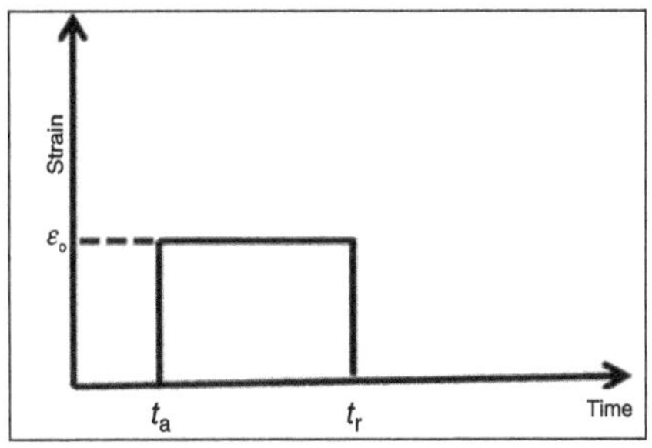

Elastic deformation.

On removal of the stress, the material recovers its original dimensions completely, that is the deformation is reversible, that is:

$$\sigma = E\varepsilon$$

Where E is the elastic modulus, σ is applied stress, and ϵ is the strain developed.

However, in viscous deformation, the strain generated is not instantaneous and it is time dependent. The strain keeps on increasing with time on application of the constant load, that is, the recovery process is delayed. When the load is removed, the material does not return to its original dimensions completely, that is, this deformation is irreversible.

$$\sigma = \gamma \frac{d\varepsilon}{dt}$$

Where,

- γ =viscosity.
- $d\epsilon/dt$=strain rate.

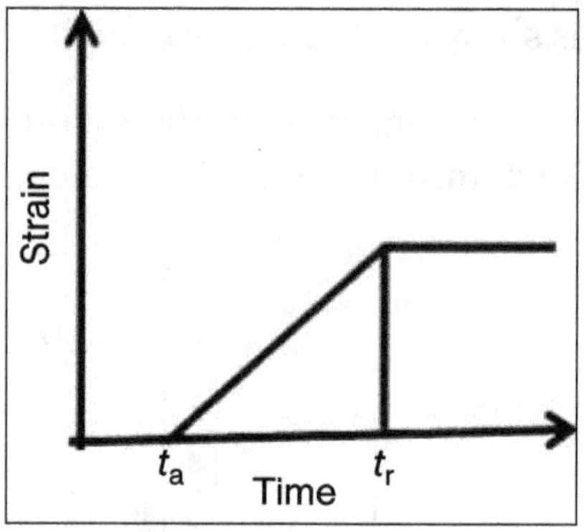

Viscous deformation.

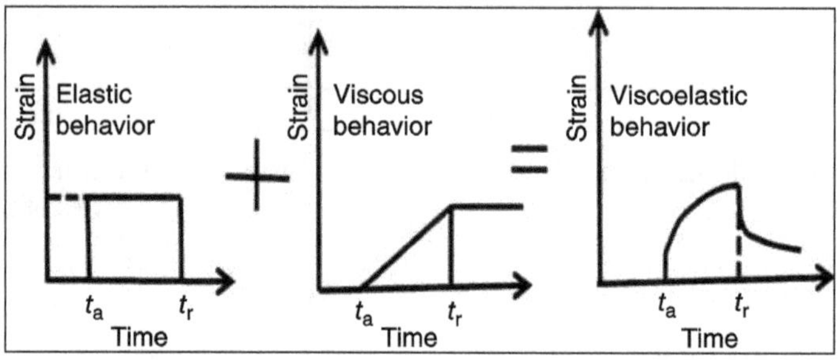

Viscoelastic deformation: the combined behavior of viscosity and elasticity.

Usually, polymers show a combined behavior of elastic and plastic deformation depending on the temperature and strain rate. At low temperature and high strain rate, elastic behavior is observed, and at high temperature but low strain rate, the viscous behavior is observed. The combined behavior of viscosity and elasticity is observed at intermediate temperature and strain rate values. This behavior is termed as viscoelasticity, and the polymer is termed as viscoelastic.

Viscoelastic Relaxation Modulus

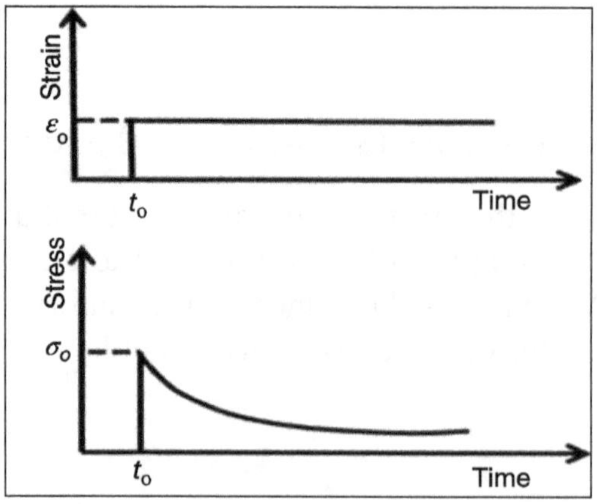

Stress relaxation in polymer.

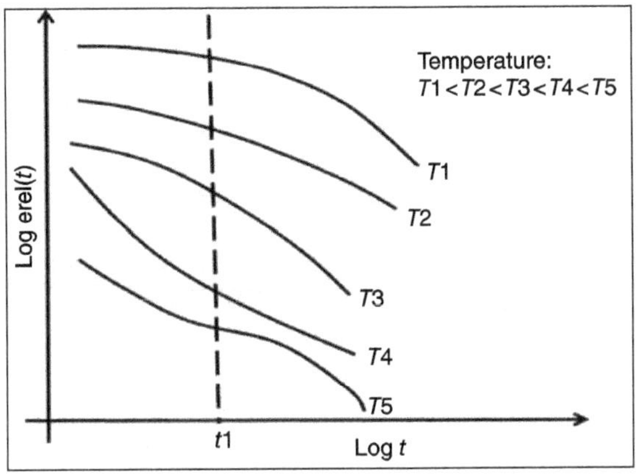

Variation of relaxation modulus with temperature and time.

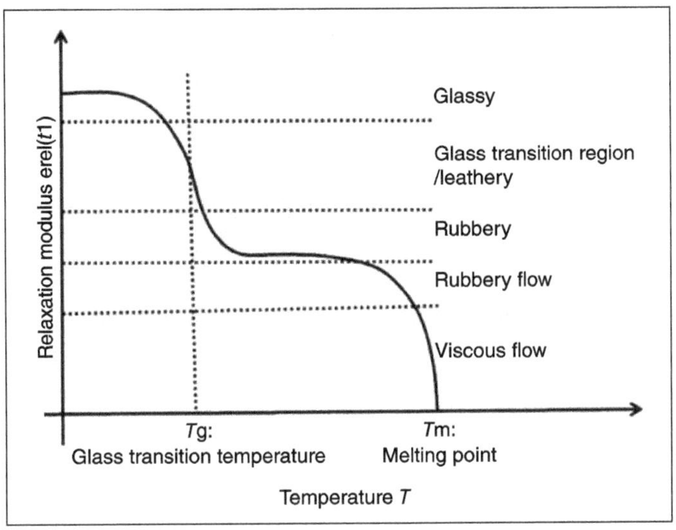

Variation of relaxation modulus with temperature after a given time t.

At a given temperature, when the polymer is strained to a given value, then the stress required to maintain this strain is found to decrease with time. This is called stress relaxation. The stress required to maintain the constant strain value is decreased with time, because the molecules of polymer get relaxed with time, and to maintain the level of strain, somewhat lower value of stress is sufficient.

$$E_{\text{rel}}(t) = \frac{\sigma(t)}{\varepsilon_0}$$

The decrease in stress follows the exponential decay:

$$\sigma = \sigma_0 e^{-t/\tau}$$

Where,

- σ =stress at time t,
- σ_o =peak stress level, and
- τ =relaxation time constant.

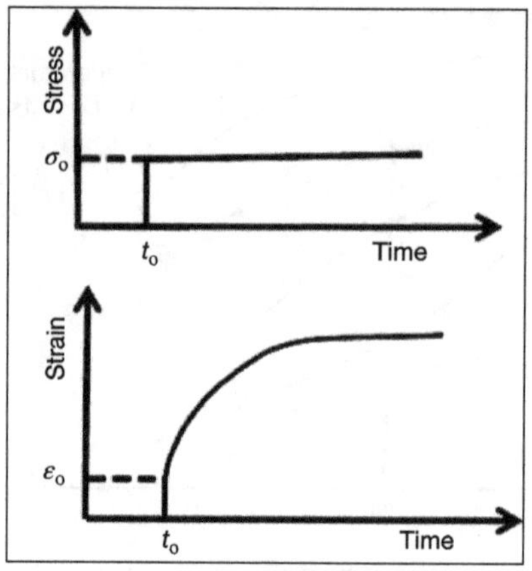

Viscoelastic creep.

The relaxation modulus is found to decrease with increase in temperature and time as shown in figure.

Now, consider the time t1. Measure the values of relaxation modulus at time t1 at different temperatures, say T1, T2, T3 ... for the viscoelastic polymer and plot the relaxation modulus versus temperature. A typical plot is shown in figure. The glass transition temperature lies near the upper temperature extremity.

Viscoelastic Creep: creep can be considered as the opposite of stress relaxation (lowering of the stress with time maintaining the constant strain level) where the polymer suffers time-dependent deformation (increasing strain with time) at constant stress level. At a given temperature, when a constant load is applied to the material, there is a time-dependent increase of strain in the material. This behavior of the material is called viscoelastic creep. The increase in molecular weight and stiffness of the chains results in better creep resistance of the material.

We can define the time-dependent creep modulus as follows:

$$E_{\text{Creep}}(t) = \frac{\sigma_0}{\varepsilon(t)}$$

Or we can define the creep compliance as follows:

$$J_{\text{Creep}}(t) = \frac{\varepsilon(t)}{\sigma_0}$$

where,

- σ_0 =constant applied stress.
- (t)=strain at time t.

Chemical Properties

The chemical properties of polymers are very similar to those of analogous small molecules. A functional group attached to a polymer chain generally reacts the same way it would if present in a monomer. For example, an acid group can be esterified, an aromatic ring can be sulfonated, and anally lie hydrogen can be abstracted by free radicals. The rates at which these pendant groups undergo reactions, however, can be quite different. For example, due to steric effects and the hydrophobicity of its surroundings, the ester group in poly- (methyl methacrylate) is considerably more resistant to hydrolysis than the ester group in methyl propionates.

Conversely, the reactivity of pendant functional groups can be substantially increased by the presence of suitable neighboring groups. For example, the rate of transesterification of pendant phenyl-ester groups surrounded by pendant carboxyl groups is many orders of magnitude faster than that of analogous phenyl acetates.

Optical Properties

Optical properties of polymers, such as, gloss, transparency, clarity, haze, colour, surface aspect and refractive index, are closely linked to our perception of a plastic product's quality and visual performance. Polymer optical properties testing will bring your insight to accelerate development and to optimise or troubleshoot production.

Optical properties are dependent on the specific polymer or copolymer material, the formulation (colourants, fillers, plasticizers and other additives) and the crystallinity of the materials. But also mechanical and chemical degradation or aging processes and the mechanical conditions and / or temperature treatments applied during processing influence the optical properties of polymers and polymer formulations. Any negative impact to a polymer's optical properties - such as aging - can affect the product's lifetime, even whilst the functional or mechanical properties still meet specifications.

The aesthetics of a product must be maintained throughout product's packaging, transport and storage, or in the automotive industry where a consumer's perception of 'glossy', strong, surface features are inextricably linked to the 'value' of the product.

From our independent perspective, we expertly analyse a range of optical properties, delivering accurate information which helps to achieve the desired aesthetics during development, production or in the end-use application.

During production, various steps such as colouration, heat treatment and mechanical processing can affect optical properties. Through our testing programs - including fundamental tests such as Gloss, Haze, Birefringence and Ultra-Violet and Visable (UV/VIS) light absorption and transmission - Intertek can help you to optimise production in order to achieve optical property specifications.

Our polymer scientists understand that the optical properties of polymers are closely related to the chemical microstructure, thermal and physical properties of these materials, and as such, providing comprehensive expertise which spans chemical, thermal plus physical and mechanical testing. Our industry application experts take you beyond the test result, applying their experience and knowledge to gain improvement of a material's optical properties.

Intertek's scientists understand that the optical properties of polymers are closely related to the chemical microstructure, thermal and physical properties of these materials, and as such, we provide comprehensive expertise that covers chemical, physical, mechanical and thermal testing. Our industry application experts take you beyond the test result, using their experience and knowledge to provide the guidance you need to achieve improvement of a material's optical properties.

Color Measurement

Most polymers are typically grey or yellow coloured. White fillers or other optically active additives colour additives can help to make the appearance more suited to application. Our experts use colour measurement via LAB colour space and XYZ colour space to evaluate color change due to exposure to adverse conditions or to help match parts molded from different materials or assess color changes due to surface texture or molding or processing.

Yellowness Index

Yellowness Index is a number calculated from spectrophotometric data that describes the change in color of a test sample from clear or white to yellow. Intertek uses the yellowness index test to evaluate color changes in a material caused by real or simulated outdoor exposure.

Specular Gloss

Gloss is the capacity of the polymer surface to reflect light in a given direction. Intertek provides testing for gloss on a range of sample types and sizes, at different incidence angles delivering comparative data between similar samples thus supporting development and end use performance testing.

Haze

Haze is an optical effect caused by light scattering within a transparent polymer resulting in a cloudy or milky appearance. Our polymer scientists use spectrophotometer techniques to investigate a range of issues which result in haze as a symptom such as weathering, during product and process development.

Refractive Index

The refractive index and light transmission of transparent materials for example, those use in lenses, optical fibres, large screen displays or coatings, are relatively straightforward to measure accurately. Intertek's polymer laboratories use refractive index as a diagnostic tool to help understand the critical stress in a system or the resulting impact of additives or heat treatments (such as cross linking, or annealing) have on the material. We also provide ultra-violet and visible (UV/VIS) light absorption and transmission measurements.

Birefringence

The refractive index varies with orientation of the polymer chains on a molecular level; our polymer scientists use polarized light to achieve birefringence effects which can be used to quantify the stress in a transparent plastic.

Electrical Properties

Materials such as glass, ceramics, polymers and bio composites are non-conducting materials. They prevent flow of current through them. When these types of non-conducting materials are placed in an electric field, they modify the electric field and they themselves undergo appreciable changes as a result of which they act as stores of electrical charges. When charge storage is the main function, the materials are called dielectrics. For a material to be a good dielectric, it must be an insulator. As good insulators, polymers possess excellent dielectric properties. Many authors have reported theoretical and experimental work related to these properties. The common electrical properties of interest are discussed below:

Dielectric Polarization

Polarization with the application of an electric field, due to the displacement of charge particles inside the material forming dipoles, is known as internal polarization. Some of the internal polarizations are electronic, orientational (dipolar), space charge and barrier polarization.

Dielectric can also be charged by direct injection of charge carriers, when high electric field ($\approx 10^5$ V/cm) is applied between the electrodes in intimate contact with it. Charges get sprayed or deposited due to corona discharge or dielectric breakdown in the thin air gap between the electrode and the dielectric. If the injected charge in the surface is of same sign as that of electrode in contact, it is called homo-charge; if it is opposite in sign then it is called hetero-charge. The total polarization is the resultant of both internal and external polarization.

$$P = P_{int} + P_{ext}$$

Polymers as dielectrics are known to store charge permanently when subjected to field - temperature treatment; such quasi permanently charged dielectrics are known as Electret.

Dielectric Strength

It measures the highest current that can be applied to a plastic before it allows current to pass. It is expressed as the voltage just before this happens divided by the thickness of the sample (in volts/m). It is affected by temperature, thickness, how the sample was conditioned, rate of voltage increase, test duration and contamination etc.

Electrical Conduction

The electrical conductivity of polymer, x, measures the presence of free ions not connected chemically with the macromolecules. It also depends on presence of low molecular weight impurities that can serve as source of ions. The chemical constitution has only an indirect effect on the mobility of the ions.

In glassy state the conductivity of the polymer is approximately 10^{-13} to 10^{-19} ohm^{-1} . With increasing temperature, the conductivity of polymer increases according to the exponential law as,

$$x = A\, e^{-\Delta U/RT}$$

where, A is a Coefficient mainly dependent on temperature, R is Universal gas constant, and ΔU is the Activation energy.

Dielectric Constant (or Permittivity)

Dielectric constant, $\varepsilon^{\prime}$, indicates how easily a polymer/plastic can be polarized relative to vacuum. It is defined as the ratio of the capacity of an electric capacitor filled with the substance to that of the same capacitor in vacuum, at a definite external field frequency. This dimensionless number which is important in high frequency applications varies with temperature, moisture, frequency and thickness.

Dissipation Factor

This measure the energy dissipated during rapid polarization reversals, as with an alternating current. It can be seen as the ratio of energy lost as heat to current transmitted. It is usually measured at 1 MHz. This factor should be low when polymers are used as insulators in high-frequency applications such as radar and microwave equipments.

Dielectric Loss

Dielectric loss, $\varepsilon^{\prime\prime}$, measures the part of the energy of an electric field that is dissipated irrecoverably as heat in the dielectric. Dielectric loss in polymeric materials is due to the independent movement of chain sections consisting of large number of monomer units.

Volume Resistance

A standard measure of conductivity when a direct current potential is applied across a material is volume resistivity (measured as ohm × area of the smaller electrode/ specimen thickness). Materials measuring volume resistance above 108 ohm-cm are insulators.

Surface Resistance

This expresses how well current flows over the surface of a material between electrodes placed on the same side of a specimen. While volume resistance is a property of the material, surface resistance measures how susceptible a plastic is to surface contamination, especially moisture. It is useful when surface leakage may be a problem but since it is not measurable exactly it should be used with wide margin of safety.

Polymer Chemistry

Polymer chemistry or macromolecular chemistry is a branch of chemistry that deals with the preparation and properties of polymers and macromolecules, and one of several fields that play an important role in polymer science and technology. Due to their commercial importance, most research in polymer chemistry is concerned with synthetic organic polymers, such as plastics or fibers. Many polymer chemists, however, work on problems related to medicine, biology and biochemistry, or materials science.

Naturally occurring polymers such as amber and rubber have been used by humans for millennia. Early Mesoamericans are perhaps the true pioneers in polymer chemistry, having discovered methods for treating natural rubber that were not reproduced until thousands of years later.

The earliest work in modern polymer chemistry involved the chemical modification of naturally occurring polymers. The reaction between nitric acid and cellulose, studied by Henri Braconnot in 1832 and later by Christian Schönbein, led to the discovery of nitrocellulose and celluloid. The ensuing years saw the preparation of other cellulose derivatives, such as collodion, used as a wound dressing since the U.S. Civil War, and cellulose acetate, first prepared in 1865.

Other early work in polymer chemistry involved the modification of natural rubber to improve durability. In 1834, Friedrich Ludersdorf and Nathaniel Hayward independently discovered that adding sulfur to raw natural rubber (polyisoprene) helped prevent the material from becoming sticky. In 1844 Charles Goodyear received a U.S. patent for vulcanizing rubber with sulfur and heat. Thomas Hancock had received a patent for the same process in the U.K. the year before.

In 1884 Hilaire de Chardonnet started the first artificial fiber plant based on regenerated cellulose, or viscose rayon, as a substitute for silk, but it was very flammable. In 1907 Leo Baekeland invented the first wholly synthetic polymer, a thermosetting phenol-formaldehyde resin called Bakelite. Cellophane was invented in 1908 by Jocques Brandenberger who squirted sheets of viscose rayon into an acid bath.

The work of Wallace Carothers in the 1930s demonstrated that polymers of desired chain length and composition could be synthesized rationally from constituent monomers, laying the foundations of modern polymer chemistry and laying the framework for the now burgeoning polymer industry. Carothers is credited with the development of neoprene, a synthetic rubber, the first polyester, and nylon, a true silk replacement. The work of Ziegler and Natta in the 1950s laid the basis for stereospecific polymer synthesis. Stephanie Kwolek developed an aramid, or aromatic nylon named Kevlar, patented in 1966.

There are now a large number of commercial polymers, including composite materials such as carbon fiber-epoxy, polystyrene-polybutadiene (HIPS), acrylonitrile-butadiene-styrene (ABS), and other such materials that combine the best properties of their various components, including polymers designed to work at high temperatures in automobile engines.

The American Chemical Society estimates that 50% of chemistry professionals will work in a polymer-related field for some portion of their career. Though polymer chemists typically earn an advanced degree in synthetic organic chemistry, some institutions offer specialized degree programs in materials science and polymer science to meet the evolving needs of the polymer industry. Given the current commercial importance of synthetic polymers, most jobs in polymer chemistry are industrial jobs.

Current areas of active interest in polymer chemistry include the following:

- Fundamental research into controlled syntheses and novel polymerization reactions.
- Development of new molecular architectures, such as supramolecular polymer complexes.
- Development of molecular architectures suited for molecular sensor technology.

- Syntheses of polymers with energy and charge transport properties.
- Biomedical applications, such as novel protein design and synthesis and targeted drug delivery.

There is also emerging interest in green polymer chemistry. Most artificial hydrocarbon-based polymers are formed from petroleum products. Substantial research efforts are devoted to improved recycling methods, renewable sources of raw materials, and biodegradable polymer materials.

Basic Concepts in Polymer Chemistry

Polymers and Macromolecules

In chemistry, the terms "polymer molecule" and "macromolecule" are used interchangeably. A polymer molecule has a high molecular mass and is comprised of many smaller, repeating subunits or *monomers*. Polymers may be found in nature, such as the DNA and proteins found in living cells, or created in laboratories or factories.

Polymer molecules come in many shapes and sizes. A polymer molecule may be a long chain of a single monomer repeated over and over again or a complex network containing dozens of different types of monomers. The identity, variety, and arrangement of monomers in a polymer molecule affect the chemical and physical properties of the polymer molecule.

Polymer Synthesis

An important area of research in polymer chemistry is finding new or better ways to prepare a polymer molecule from a stock of smaller monomers. In most cases, polymers are prepared using principles of organic chemistry. Polymer chemists are especially interested in techniques that allow them to precisely control the size and structure of the end product.

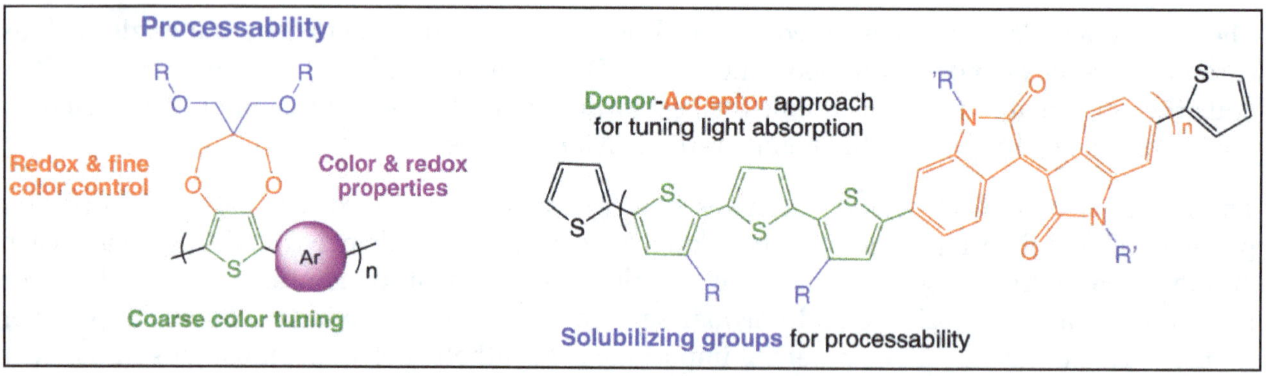

Polymer chemists are also investigating polymerization methods outside the scope of organic chemistry. One area of interest involves preparing polymers by imitating the biological processes used to create biopolymers such as proteins or cellulose. Other areas of study involve using plasma or electricity to initiate polymerization reactions.

Physical Polymer Chemistry

Physical polymer chemistry is the study of how a polymer molecule's structure relates to the behavior of the bulk substance. Physical polymer chemistry is closely related to the field of polymer physics and also overlaps with polymer research in materials science. Physical polymer chemists use analytical techniques such as light scattering and spectroscopy to characterize the size and structure of polymers.

Other areas of interest in physical polymer chemistry include the study of polymers in solution, the mechanical properties of polymers, and understanding phase transitions in polymer substances. There are also many researchers using principles of theoretical chemistry to better understand the structure and properties of polymer molecules.

2

Classification of Polymers

Polymers are classified on the basis of source, structure, mode of polymerization, molecular forces, etc. It includes rubber polymers, biological polymers, synthetic polymers, linear polymers, addition polymers, elastomers, etc. This chapter has been carefully written to provide an easy understanding of these different types of polymers.

Classification based on Source

Naturally Occurring Polymers

These polymers are found in nature generally from plants and animals sources. Such as proteins, cellulose, starch, resins, etc.

One reason why natural polymers have become an interest in undoubtedly because it can be obtained from natural sources that are abundant and renewable also ease to get. Polysaccharides are found in abundance in nature and are readily available from sources such as algae (e.g. alginates), plants (e.g. pectin, guar gum, mannan), microbes (e.g. dextran, xanthan gum) and animals (e.g. chitosan, chondroitin) and they can also be produced by means of recombinant DNA techniques. Monosaccharide polymers have many favourable properties such as high stability, nontoxicity, hydrophilicity, biodegradability, gel forming properties and ease of chemical modification. An enormous variety in plant polysaccharide structural composition exists, which is not only associated with different plants, but also with the part of the plant that they originate from, such as the leaves, seeds, roots and tubers. The complexity and variety of polysaccharides can be explained by two unique structural features: firstly monosaccharides can be linked together in different ways and in an α- or β configuration and secondly, due to the presence of branched side-chains.

Physicochemical Properties of Natural Polymers

Homogenecity/Polydispersity

Natural polymers such as proteins are rather uniform in size and there is no variation. Such polymers are said to be homogeneous or monodisperse. Most of natural polymers are naturally built by condensation polymerization. Natural polymers tend to be readily biodegradable and they show no adverse effects on the environment or human beings.

Rubber Polymers

Rubber is a natural polymer of Isoprene (2-Methyl -1, 3 – Butadiene). It is a linear, 1, 4 – addition polymer of Isoprene.

Natural rubber has elastic properties and it undergoes long range reversible extension even if relatively small force is applied to it. Therefore, it is also known as Elastomer. Natural rubber is prepared from latex which is a Colloidal Solution of Rubber in Water.

Rubber trees are basically found in tropical & semitropical countries. Indonesia, Malaysia, Sri Lanka, South America and India (especially Kerala, Tamilnadu and Karnataka) have abundant resource of natural rubber.

Rubber Plantation in Kerala.

Structure of Natural Rubber

Natural Rubber is a polymer of Isoprene. To understand the structure of Rubber we shall concentrate on structure of Isoprene. Isoprene is a conjugated diene containing double bonds at alternate position.

$$\begin{array}{c} \quad CH_3 \\ \quad | \\ CH_2 = C - CH = CH_2 \end{array}$$

Structure of Isoprene: Monomer of Natural Rubber.

Isoprene undergoes free radical polymerization like substituted ethylene. Isoprene polymerizes to give Polyisoprene polymer, a simple alkene having each unit still containing one double bond.

Now, polymerization of Isoprene may follow either of the two pathways; either of cis-polymerization or trans-polymerization. The rubber formed from cis-polymerization is called cis-Polyisoprene or Natural Rubber. Similarly, the rubber formed from trans-polymerization is called Synthetic Rubber.

$$n\,CH_2 = \overset{\overset{\displaystyle CH_3}{|}}{C} - CH = CH_2 \longrightarrow \left[CH_2 - \overset{\overset{\displaystyle CH_3}{|}}{C} = CH - CH_2 \right]_n$$

Isoprene (2-Methyl -1, 3-butadiene) undergoes cis-polymerization to form natural rubber.

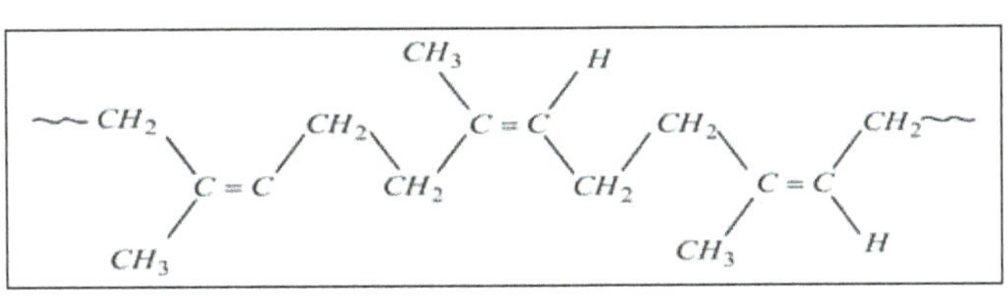

Structure of Natural Rubber (Cis Polyisoprene)

By observing structure of natural rubber we can infer that there is no polar group in this structure. As a result of this the intermolecular forces of attraction are weak Vanderwaal forces of attraction. These forces of attraction are further weakened because of the cis-configuration of all the double bonds that does not permit the close interaction of polymer chains. Thus Natural Rubber (Cis Polyisoprene) does not have a straight chain but has a coiled structure. As a result of this, it gets elastic property.

Synthetic Rubber

Polymerization of deines (molecules containing double bond) to form substitutes for rubber is the forerunner of the enormous present day plastic industry. Chloroprene was the first commercially successful rubber substitute produced in the United States.

$$CH_2 = \overset{\overset{Cl}{|}}{C} - CH = CH_2 \xrightarrow{\text{Polymerisation}} \left(CH_2 - \overset{\overset{Cl}{|}}{C} = CH - CH_2 \right)_n$$

Chloroprene undergoing trans-polymerization to produce Synthetic Rubber, Polychloroprene.

The properties of Rubber so formed are determined by the nature of the substitutent groups. For example, Polychloroprene is inferior to natural rubber in some properties but superior in its resistance to oil, Organic Solvents. These differences are due to difference in nature of their monomers: Isoprene (for natural rubber) and Chloroprene (for synthetic rubber).

Synthetic Rubber (also known as Gutta-Percha) was obtained by the free radical polymerization of Isoprene. The rubber so formed has all trans- Configuration. As a result of this, synthetic rubber has a highly regular zig-zag chain which cannot be stretched .This accounts for non-elasticity of Synthetic Rubber.

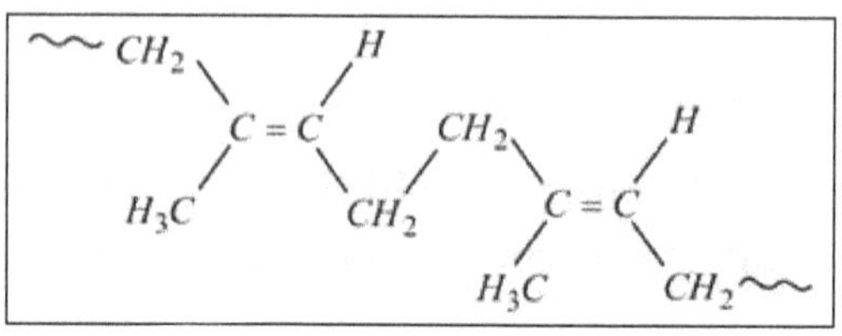

Structure of Synthetic Rubber (trans – Polyisoprene, Gutta-percha).

Types of Synthetic Rubber

Neoprene

Neoprene is a polymer of chloroprene. It is also known as Polychloroprene. To synthesize Neoprene its monomer Chloroprene is required. Chloroprene required for this process is synthesized from Vinylacetylene which performs Markonikov addition under acidic condition to produce Chloroprene.

$$CH_2 = CH - C \equiv CH \; + HCl \xrightarrow[\text{Addition}]{\text{Mark}} CH_2 = CH - \underset{\underset{Cl}{|}}{C} = CH_2$$

(Chloroprene)

The Vinylacetylene required for above reaction is prepared by Dimerization of acetylene by passing it through an aqueous solution of Ammonium Chloride and Cuprous Chloride at 343K.

$$\underset{\text{Acetylene (2molecules)}}{CH \equiv CH + HC \equiv CH} \xrightarrow[cucl,343K]{NH_4cl,} CH_2 = CH - CH$$

The Chloroprene obtained undergoes Polymerization to gives Neoprene. Though no specific catalysts are needed for this process but the polymerization becomes faster in the presence of Oxygen or peroxide.

$$\underset{\text{(Chloroprene)}}{n\, CH_2 = \underset{cl}{\underset{|}{C}} - CH = CH_2} \xrightarrow[peroxide]{O_2 or} \underset{\text{Neoprene}}{\left(CH_2 - \underset{cl}{\underset{|}{C}} = CH - CH_2 \right)_n}$$

Uses of Neoprene

It is used in the manufacture of hoses, gaskets, shoe heels, stoppers, conveyor belts and printing rollers etc. It is also used as an insulator.

BUNA–S

BUNA–S is a copolymer of a mixture of 1,3- Butadiene and styrene in the ratio of 3:1 in the presence of sodium (which is polymerizing agent) gives styrene – butadiene copolymer (styrene – butadiene rubber) or BUNA –S. The name BUNA–S is made up of Bu which indicates 1, 3 – Butadiene, NA is for Sodium (Na) and S indicates Styrene.

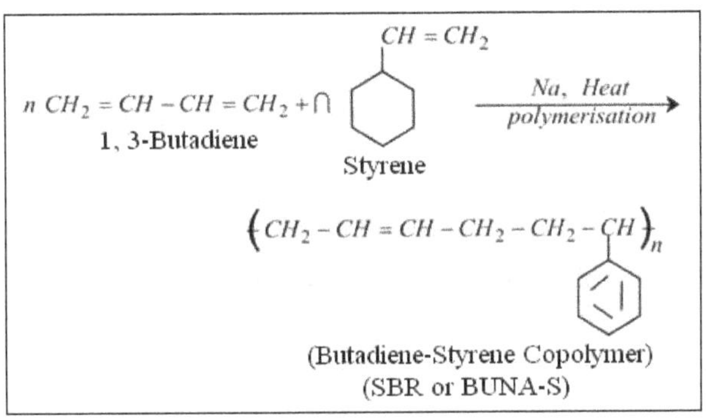

Properties of Buna–S

- It is very tough and a good substitute for natural rubber.
- It possesses high abrasion resistance.
- It has high load bearing capacity.

Uses of Buna–S

- It is used for manufacturing automobile tyres.
- It is used for making floor tiles, footwear components, cable insulation etc.

BUNA – N (Nitrile Rubber)

BUNA–N is obtained by copolymerization of 1, 3 – Butadiene and acrylonitrile in presence of a peroxide catalyst. The name BUNA–N is made up of Bu which indicates 1, 3 – Butadiene, NA is for Sodium (Na) and N indicates acrylonitrile.

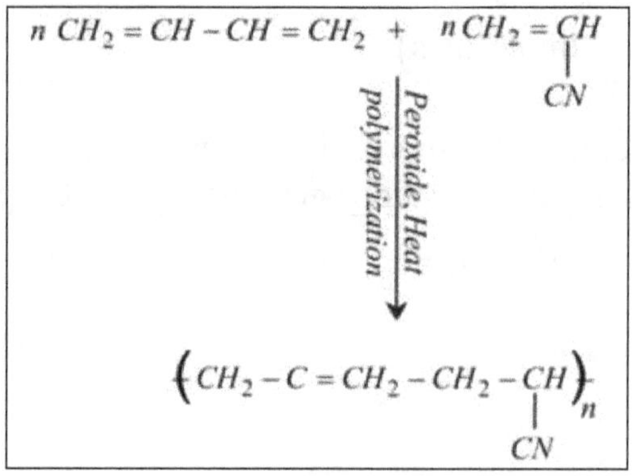

1, 3-Butadiene reacts with Acrylonitrile to produce Buna – N.

Properties of BUNA–N

BUNA-N is resistant to the action of petrol, lubricating oils and organic solvents.

Uses of BUNA–N

BUNA-N It is used in making oil seals, hoses, tank linings etc.

Thiokol

Thiokol is prepared by copolymerization of 1, 2 – dichloroethane (ethylene dichloride) with Sodium Tetrasulphide (Na_2S_4) in presence of Magnesium hydroxide.

Properties of Thiokol

Thiokol is resistant to the action of mineral oils, solvents, oxygen & ozone.

Vulcanized Rubber

Natural rubber is not an important polymer for commercial purpose because of its softness & tacky (sticky) properties. Softness of natural rubber increases with the increase in temperature while brittleness increases at low temperature.

Therefore, ideal temperature for using Rubber is 283 – 335K where its elasticity is maintained. Other properties which decrease the quality of natural rubber are:

- It has large water absorption capacity.
- It has low tensile strength and low resistance to abrasion.
- It is not – resistant to abrasion.
- It is easily attacked by organic reagents.

The properties of Natural rubber can be improved by a process called Vulcanization .Vulcanization is the process of introduction of Sulphur bridges between different chains by heating raw rubber with Sulphur at 373-415K. In the absence of catalyst the process of vulcanization is slow. Some additive such as Zinc Oxide is added to accelerate the rate of Vulcanization.

Difference between Vulcanized Rubber and Natural Rubber

The new or vulcanized rubber obtained has properties that are just opposite to that of natural rubber. These properties are:

- Vulcanized rubber has excellent elasticity.
- Low water absorption tendency.
- It is resistant to the action of organic solvents.
- It is resistant to attack of oxidizing agents.

Vulcanized Rubber is an improved form of Natural rubber. In vulcanized rubber, Sulphur bridges are introduced either at their reactive allylic sites or at the site of double bond. The presence of double bond in the rubber molecule makes it's highly reactive as it provides allylic hydrogen that permits formation of Cross links between different chains. The presence of these cross links increases the toughness, strength and hardness of rubber. Due to the presence of Sulphur bridges, individual chains can no longer slip over one another but are locked together in a giant size molecule.

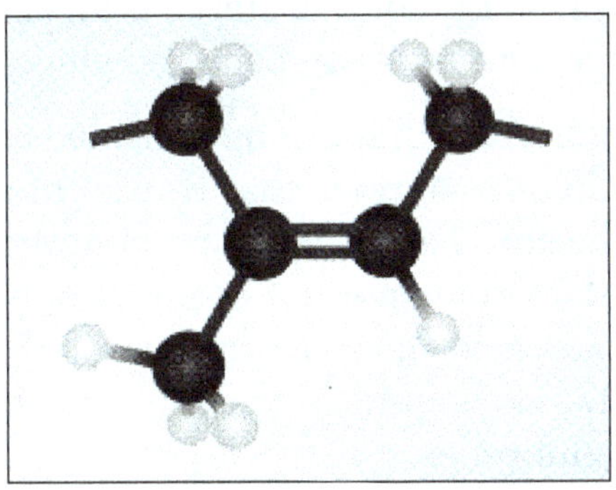

Cross-links being formed between the rubber polymer chains during Vulcanization.

Biological Polymers

Biological polymers are large molecules composed of many similar smaller molecules linked together in a chain-like fashion. The individual smaller molecules are called monomers. When small organic molecules are joined together, they can form giant molecules or polymers. These giant molecules are also called macromolecules. Natural polymers are used to build tissue and other components in living organisms.

Generally speaking, all macromolecules are produced from a small set of about 50 monomers. Different macromolecules vary because of the arrangement of these monomers. By varying the sequence, an incredibly large variety of macromolecules can be produced. While polymers are responsible for the molecular "uniqueness" of an organism, the common monomers mentioned above are nearly universal.

The variation in the form of macromolecules is largely responsible for molecular diversity. Much of the variation that occurs both within an organism and among organisms can ultimately be traced to differences in macromolecules. Macromolecules can vary from cell to cell in the same organism, as well as from one species to the next.

Biomolecules

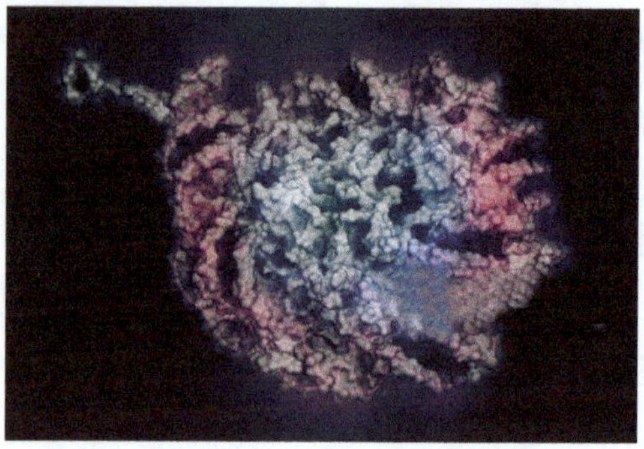

There are four basic kinds of biological macromolecules. They are carbohydrates, lipids, proteins, and nucleic acids. These polymers are composed of different monomers and serve different functions.

- Carbohydrates - molecules composed of sugar monomers. They are necessary for energy storage. Carbohydrates are also called saccharides and their monomers are called monosaccharides. Glucose is an important monosaccharide that is broken down during cellular respiration to be used as an energy source. Starch is an example of a polysaccharide (many saccharides linked together) and is a form of stored glucose in plants.
- Lipids - water-insoluble molecules that can be classified as fats, phospholipids, waxes, and steroids. Fatty acids are lipid monomers that consist of a hydrocarbon chain with a carboxyl group attached at the end. Fatty acids form complex polymers such as triglycerides, phospholipids, and waxes. Steroids are not considered true lipid polymers because their molecules do not form a fatty acid chain. Instead, steroids are composed of four fused carbon ring-like structures. Lipids help to store energy, cushion and protect organs, insulate the body, and form cell membranes.
- Proteins - biomolecules capable of forming complex structures. Proteins are composed of amino acid monomers and have a wide variety of functions including transportation of molecules and muscle movement. Collagen, hemoglobin, antibodies, and enzymes are examples of proteins.
- Nucleic Acids - molecules consisting of nucleotide monomers linked together to form polynucleotide chains. DNA and RNA are examples of nucleic acids. These molecules contain instructions for protein synthesis and allow organisms to transfer genetic information from one generation to the next.

Assembling and Disassembling Polymers

While there is variation among the types of biological polymers found in different organisms, the chemical mechanisms for assembling and disassembling them are largely the same across organisms. Monomers are generally linked together through a process called dehydration synthesis, while polymers are disassembled through a process called hydrolysis. Both of these chemical reactions involve water. In dehydration synthesis, bonds are formed linking monomers together while losing water molecules. In hydrolysis, the water interacts with a polymer causing bonds that link monomers to each other to be broken.

Cellulose

Cellulose is an organic polysaccharide located within the fiber walls of plants. This natural polymer is the most important and fascinating biopolymer.

Cellulose is insoluble in water and most common solvents, used in a wide range of applications including composites, netting, upholstery, coatings, packing, paper, etc. Many parallel cellulose molecules form crystalline microfibrils that are mechanically strong and highly resistant to enzymatic attack. These are aligned with each other to provide structure to the cell wall.

Chemical modification of cellulose is performed to improve processability and to produce cellulose derivatives (cellulosics) as carboxymethyl cellulose (CMC), methyl cellulose (MC), hydroxyethylcellulose (HEC), etc which can be tailored for specific industrial applications.

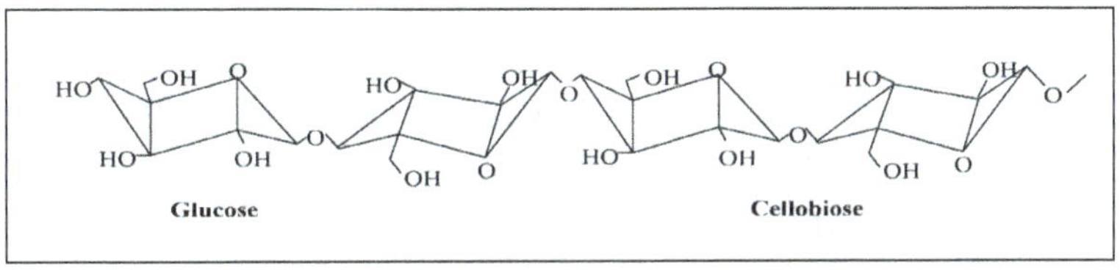

Molecular chain structure of cellulose.

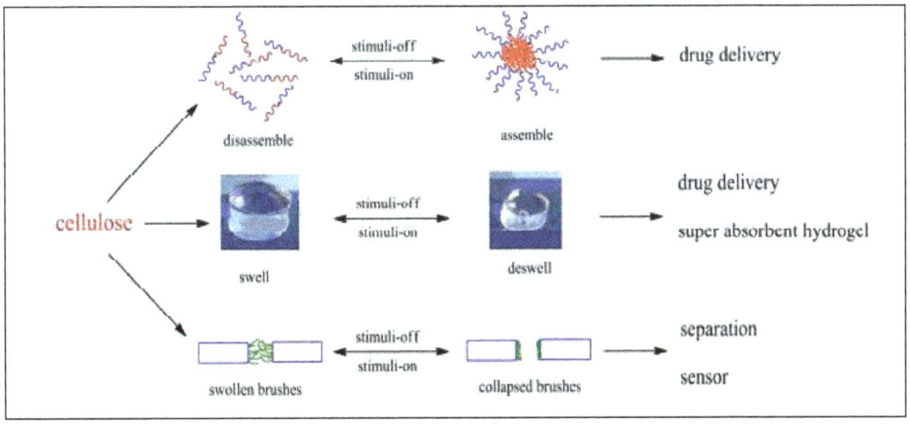

Examples of "smart" materials based on cellulose and their possible applications.

Cellulose presents many application fields as cosmetic, medical, hygiene, food and drug industry. The intramolecular bonds give stiffness to the polymer chain, while the intermolecular bonds allow the linear polymers to form sheet structures. Because this biopolymer is abundant in nature, biodegradable and cheap, it can make it a promising nano-scale reinforcement material for polymers.

Chitin

Chitin is a natural polysaccharide of great importance. This biopolymer is synthesized by enormous number of living organisms and it belongs to the most abundant natural polymers, after cellulose.

Chitin is found in the lobsters, shells of crabs, insects and shrimps or can be generated via fungal fermentation processes.This biodegradable polymer occurs in nature as ordered crystalline microfibrils forming structural components in the exoskeleton of arthropods or in the cell walls of fungi and yeast.

The semi-crystalline structure of chitin microfibrils can be treated with acid to produce whisker-shaped nanofillers which can be incorporated into polymers to elaborate nano-hybrid materials.

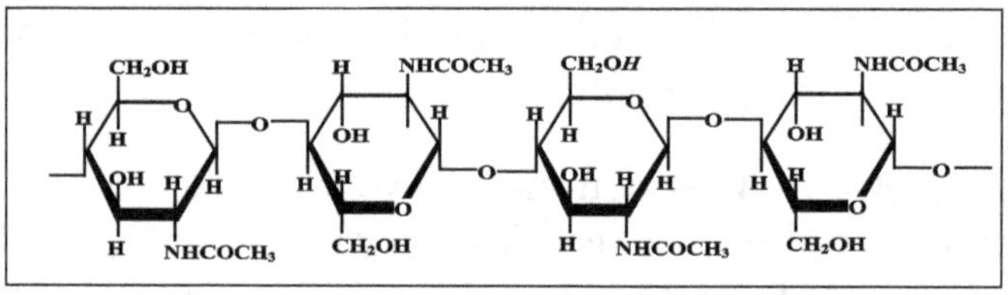

Structure of Chitin.

Physical properties of chitin in solution cannot be analyzed correctly due to poor data caused mainly by difficulties in dissolution of this polymer. The dissolution is desired to estimate the molecular weight and to process chitin. Presence of aggregates in solution precludes light scattering measurements and overestimates the molecular weights.

Many properties of chitin have been already described such as: biocompatibility, renewable origin, non-toxicity non-allergenicity and biodegradability in the body. In addition, due to their attractive biological activities such as antifungal, antibacterial, antitumor, immunoadjuvant, antithrombogenic, anticholesteremic agent and bioadhesivity.

Chitosan

Chitosan is a polysaccharide obtained from chitin by deacetylation. This biopolymer is nontoxic and inexpensive and possesses reactive amino groups that has shown to be useful in different areas as an antimicrobial compound in agriculture, as a potential elicitor of plant defence responses, as an additive in food industry, as a hydrating agent in cosmetics, as a flocculating agent in wastewater treatment and more recently as a pharmaceutical agent in biomedicine.

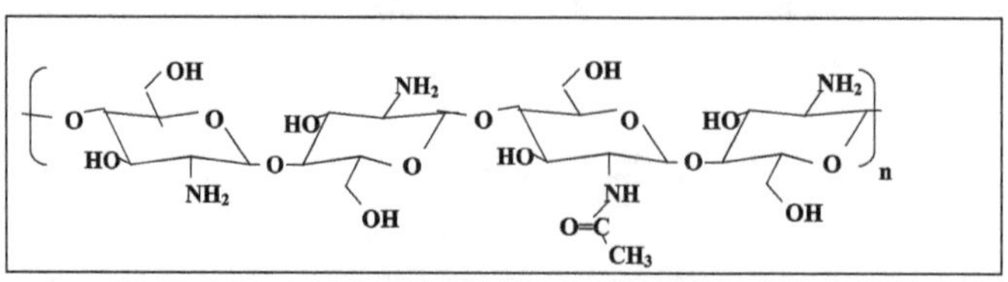

FChitosan chemical structure.

For its biocompatibility chitosan is used in various medical applications such as implantation, antacid and antiulcer activities, potential carrier for drugs, biomedical applications and it also has antibacterial action.

Chitosan can selectively bind desirable materials such as cholesterol, fats, metal ions, proteins and tumor cells and it shown affinity for proteins, such as wheat germ agglutinin and trypsin.

Gelatin

Gelatin is a biomaterial based on animal proteins, it is a pure protein food ingredient prepared by the thermal denaturation of collagen. It appears slightly yellow and commercially solid as transparent, odorless, brittle and tasteless granule, sheets, flakes or powder, soluble in hot water, glycerol and acetic acid, and insoluble in organic solvents.

This natural polymer is derived from collagen and mostly found in fibrous tissues such as tendon, ligament, and skin. It is also abundant in cornea, cartilage, bone, blood vessels, gut, and intervertebral disc.

This natural polymer has been used for long years ago in pharmaceutical formulation, cell culture and tissue engineering due to its excellent biocompatibility, ease of processing and availability at low cost.

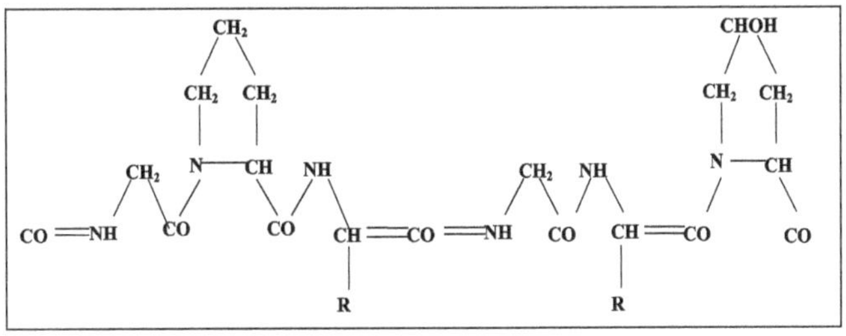

Structure of Gelatin.

Starch

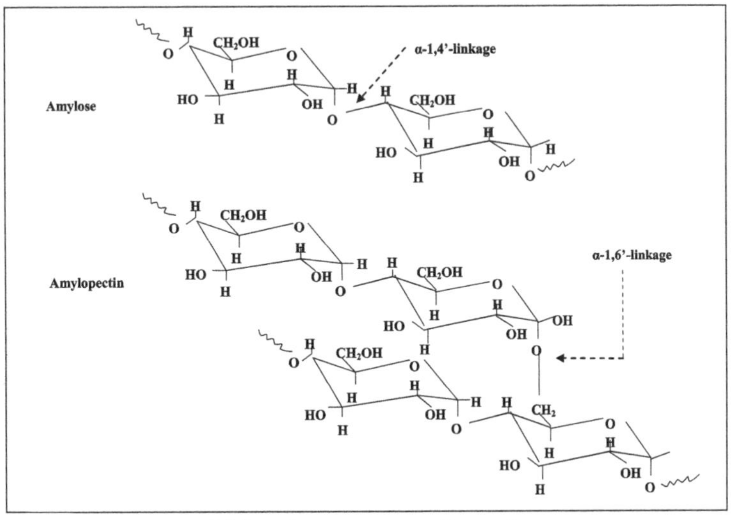

Molecular structure of starch.

Starch is a biopolymer which possesses many properties. It can be obtained from wheat, tapioca, maize and potatoes. Because it is a biodegradable polymer with well-defined chemical properties, it has a huge potential as a versatile renewable resource for various material applications in various areas. Starch can be modified in such a method that it can be melted or deformed as thermoplastics.

Starch is mainly composed of two homopolymers of D-glucose : amylase, a mostly linear α- D(1, 4')-glucan and branched amylopectin, having the same backbone structure as amylose but with many α-1, 6'-linked branch points.

This biopolymer is the major carbohydrate reserve in plant tubers and seed endosperm where it is found as granules.

Carrageenan

Carrageenan is a polysaccharide prepared by alkaline extraction from red seaweed (Rhodophycae), mostly of genus Chondrus, Eucheuma, Gigartina and Iridaea.

Carrageenans are linear polymers of about 25,000 galactose derivatives with regular but imprecise structures, dependent on the source and extraction conditions. They are composed of alternating 3-linked -d-galactopyranose (G-units) and 4-linked -d-galactopyranose (D-units) or 4- linked 3,6-anhydrogalactose (DA-units), forming the ™idealʃ disaccharide-repeating unit of carrageenans.

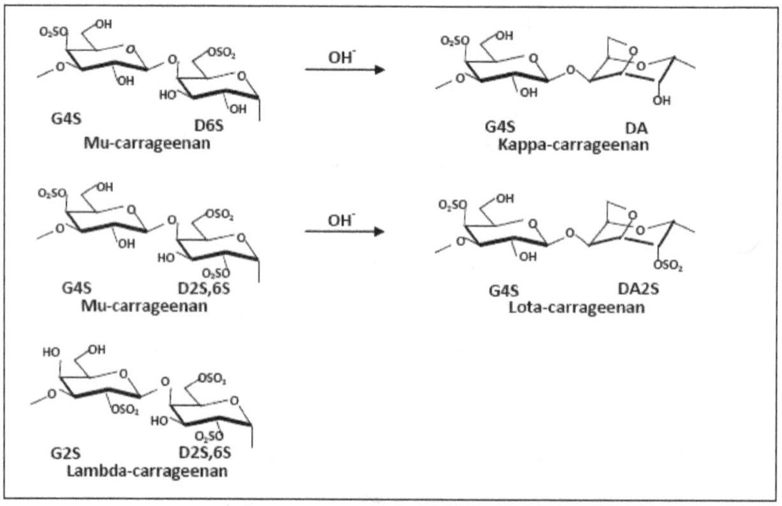

Schematic representation of the different idealized repeating units of carrageenans.
The letter codes refer to the alternative nomenclature.

Carrageenans are widely used in the food industry, for their gelling, thickening, and stabilizing properties. Their main application is in dairy and meat products, due to their strong binding to food proteins. There are three main varieties of carrageenan, which differ in their degree of sulphation. Kappa-carrageenan has one sulphate group per disaccharide. Iota-carrageenan has two sulphates per disaccharide. Lambda carrageenan has three sulphates per disaccharide.

Xylan

Xylan is a biopolymer extracted from corn cobs. This hemicellulose largely found in nature has been used as an adhesive, thickener, and additive to plastics. It increases their stretch and breaking resistance as well as their susceptibility to biodegradation.

Xylan may also be found in the food industry as an emulsifier and protein foam stabilizer during heating.

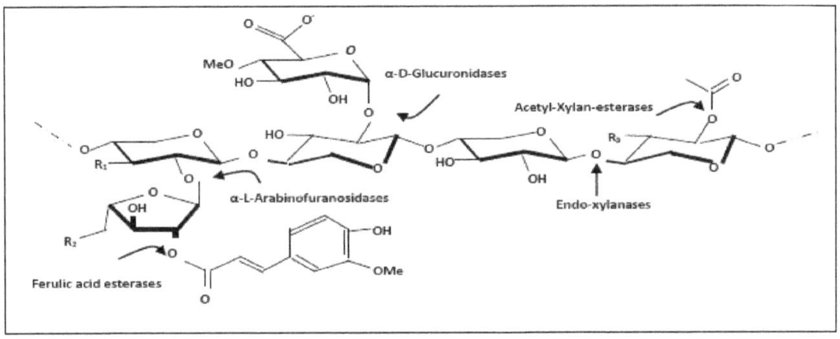

Molecular structure of xylan.

Xylans can be used as pure polymers or relatively unpurified extracts containing contaminating and/or bound protein, phenolic substances, and other polysaccharides, that is dictated by the expected end uses as well as the production costs. As the most important application fields, the pulp and paper, food and pharmaceutical industries can be marked.

Dextran

Dextran is a complex, branched polysaccharide made of many glucose molecules composed of chains of varying length (from 10 to 150 kilodaltons). This class of polysaccharides is composed of monomers of the simple sugar glucose and is stored as fuel in yeasts and bacteria.

Dextrans may be used in many important fields because they present natural products from renewable resources, excellent biocompatibility and clinical safety record and superior moisturizing properties.

The potential application of Dextran in ice-cream has been investigated. Recent studies have shown that glucose uptake by brush-border mucosa is enhanced after exposure to Dextran.

This natural polymer has been known for his viscosifying, emulsifying, texturizing, stabilizing attributes in food applications. It has the potential to be recruited as a novel ingredient replacing the commercial hydrocolloids in bakery and other food industries.

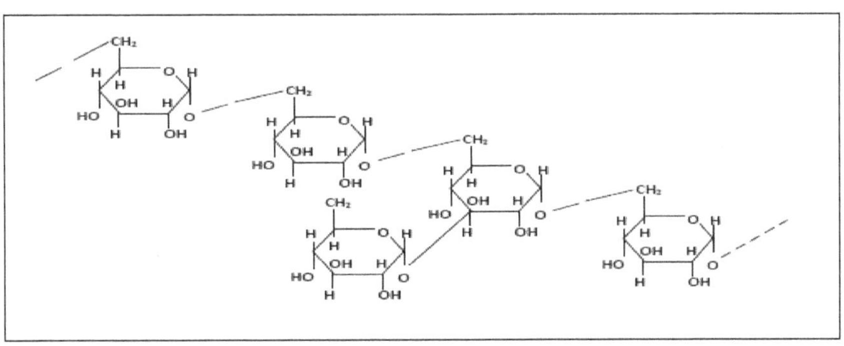

Structure of Dextran.

Synthetic Polymers

Synthetic polymers are those which are human-made polymers. Polymers are those which consists of repeated structural units known as monomers. Polyethylene is considered to be as one of the simplest polymer, it has ethene or ethylene as the monomer unit whereas the linear polymer is known as the high density polyethylene-HDPE. Many of the polymeric materials have chain-like structures which resemble polyethylene.

Synthetic polymers.

Synthetic polymers are sometimes referred as “plastics”, of which the well-known ones are nylon and polyethylene. The polymers which are formed by linking monomer units, without the any change of material, are known to as addition polymers or also called as chain-growth polymers. All these are said to be synthetic polymers.

Some of the synthetic polymers which we use in our everyday life include nylons used in fabrics and textiles, Teflon used in non-stick pans, polyvinyl chloride used in pipes. The PET bottles we use are commonly made up of synthetic polymer called as polyethylene terephthalate. The covers and plastic kits comprises of synthetic polymers such as polythene, and the tyres of vehicles are manufactured from the Buna rubbers. But on the other side, there also arises environmental issues by the use of these synthetic polymers such as the bioplastics and those made from petroleum as they are said to be non-biodegradable.

Uses of Synthetic Polymers

- The polymer called Polyethylene is used in plastic bags and film wraps.
- Polyethylene is utilized in the bottles, electrical insulation, toys, etc.
- Polyvinyl Chloride (PVC) is used in siding, pipes, flooring purposes.
- The synthetic polymer Polystyrene is used in cabinets and in packaging.
- Polyvinyl acetate is used in adhesives and latex paints.

Polyethylene

Polyethylene (PE) is light, versatile synthetic resin made from the polymerization of ethylene. Polyethylene is a member of the important family of polyolefin resins. It is the most widely used plastic in the world, being made into products ranging from clear food wrap and shopping bags to detergent bottles and automobile fuel tanks. It can also be slit or spun into synthetic fibres or modified to take on the elastic properties of a rubber.

Chemical Composition and Molecular Structure

Ethylene (C_2H_4) is a gaseous hydrocarbon commonly produced by the cracking of ethane, which in turn is a major constituent of natural gas or can be distilled from petroleum. Ethylene molecules are essentially composed of two methylene units (CH_2) linked together by a double bond between the carbon atoms—a structure represented by the formula $CH_2{=}CH_2$. Under the influence of polymerization catalysts, the double bond can be broken and the resultant extra single bond used to link to a carbon atom in another ethylene molecule. Thus, made into the repeating unit of a large, polymeric (multiple-unit) molecule, ethylene has the following chemical structure:

$$-\!\!\left[CH_2 - CH_2 \right]\!\!-$$

This simple structure, repeated thousands of times in a single molecule, is the key to the properties of polyethylene. The long, chainlike molecules, in which hydrogen atoms are connected to a carbon backbone, can be produced in linear or branched forms. Branched versions are known as low-density polyethylene (LDPE) or linear low-density polyethylene (LLDPE); linear versions are known as high-density polyethylene (HDPE) and ultrahigh-molecular-weight polyethylene (UHMWPE).

The basic polyethylene composition can be modified by the inclusion of other elements or chemical groups, as in the case of chlorinated and chlorosulfonated polyethylene. In addition, ethylene can be copolymerized with other monomers such as vinyl acetate or propylene to produce a number of ethylene copolymers.

Low-density polyethylene was first produced in 1933 in England by Imperial Chemical Industries Ltd. (ICI) during studies of the effects of extremely high pressures on the polymerization of polyethylene. ICI was granted a patent on its process in 1937 and began commercial production in 1939. It was first used during World War II as an insulator for radar cables.

In 1930 Carl Shipp Marvel, an American chemist working at E.I. du Pont de Nemours & Company (now DuPont Company), discovered a high-density material, but the company failed to recognize the potential of the product. It was left to Karl Ziegler of the Max Planck Institute for Coal Research at Mülheim an der Ruhr, W.Ger., to gain credit for inventing linear HDPE—which Ziegler actually produced with Erhard Holzkamp in 1953, catalyzing the reaction at low pressure with an organometallic compound. The process was later improved by the Italian chemist Giulio Natta, and the compounds are now known as Ziegler-Natta catalysts. In part for this innovation, Ziegler was awarded the Nobel Prize for Chemistry in 1963. Since that time, by using different catalysts and polymerization methods, scientists have produced polyethylene with various properties and structures. LLDPE, for example, was introduced by the Phillips Petroleum Company in 1968.

Major Polyethylene Compounds

Low-density Polyethylene

LDPE is prepared from gaseous ethylene under very high pressures (up to about 350 megapascals, or 50,000 pounds per square inch) and high temperatures (up to about 350 °C [660 °F]) in the presence of oxide initiators. These processes yield a polymer structure with both long and short branches. Because the branches prevent the polyethylene molecules from packing closely together in hard, stiff, crystalline arrangements, LDPE is a very flexible material. Its melting point is approximately 110 °C (230 °F). Principal uses are in packaging film, trash and grocery bags, agricultural mulch, wire and cable insulation, squeeze bottles, toys, and housewares. The plastic recycling code of LDPE is #4.

Linear Low-density Polyethylene

LLDPE is structurally similar to LDPE. It is made by copolymerizing ethylene with 1-butene and smaller amounts of 1-hexene and 1-octene, using Ziegler-Natta or metallocene catalysts. The resultant structure has a linear backbone, but it has short, uniform branches that, like the longer branches of LDPE, prevent the polymer chains from packing closely together. Overall, LLDPE has similar properties to LDPE and competes for the same markets. The main advantages of LLDPE are that the polymerization conditions are less energy-intensive and that the polymer's properties may be altered by varying the type and amount of its chemical ingredients. The plastic recycling code of LLDPE is #4.

High-density Polyethylene

HDPE is manufactured at low temperatures and pressures, using Ziegler-Natta and metallocene catalysts or activated chromium oxide (known as a Phillips catalyst). The lack of branches in its structure allows the polymer chains to pack closely together, resulting in a dense, highly crystalline material of high strength and moderate stiffness. With a melting point more than 20 °C (36 °F) higher than LDPE, it can withstand repeated exposure to 120 °C (250 °F) so that it can be sterilized. Products include blow-molded bottles for milk and household cleaners; blow-extruded grocery bags, construction film, and agricultural mulch; and injection-molded pails, caps, appliance housings, and toys. The plastic recycling code number of HDPE is #2.

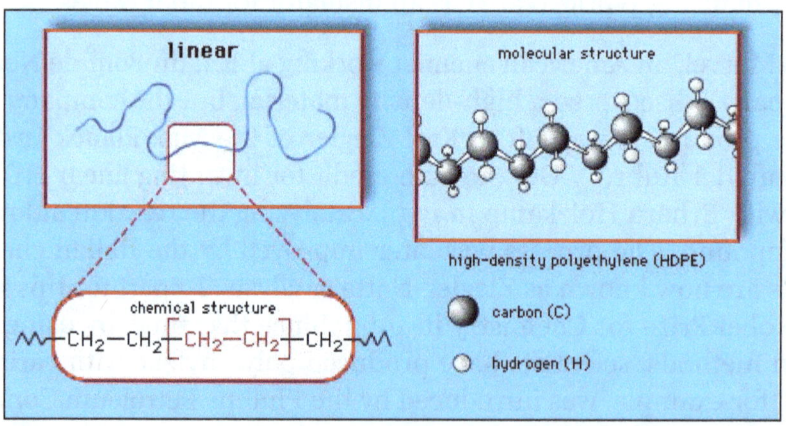

The linear form of polyethylene, known as high-density polyethylene (HDPE).

Ultrahigh-molecular-weight Polyethylene

Linear polyethylene can be produced in ultrahigh-molecular-weight versions, with molecular weights of 3,000,000 to 6,000,000 atomic units, as opposed to 500,000 atomic units for HDPE. These polymers can be spun into fibres and then drawn, or stretched, into a highly crystalline state, resulting in high stiffness and a tensile strength many times that of steel. Yarns made from these fibres are woven into bulletproof vests.

Ethylene Copolymers

Ethylene can be copolymerized with a number of other compounds. Ethylene-vinyl acetate copolymer (EVA), for instance, is produced by the copolymerization of ethylene and vinyl acetate under pressure, using free-radical catalysts. Many different grades are manufactured, with the vinyl acetate content varying from 5 to 50 percent by weight. EVA copolymers are more permeable to gases and moisture than polyethylene, but they are less crystalline and more transparent, and they exhibit better oil and grease resistance. Principal uses are in packaging film, adhesives, toys, tubing, gaskets, wire coatings, drum liners, and carpet backing.

Ethylene-acrylic acid and ethylene-methacrylic acid copolymers are prepared by suspension or emulsion polymerization, using free-radical catalysts. The acrylic acid and methacrylic acid repeating units, making up 5 to 20 percent of the copolymers, have the following structures:

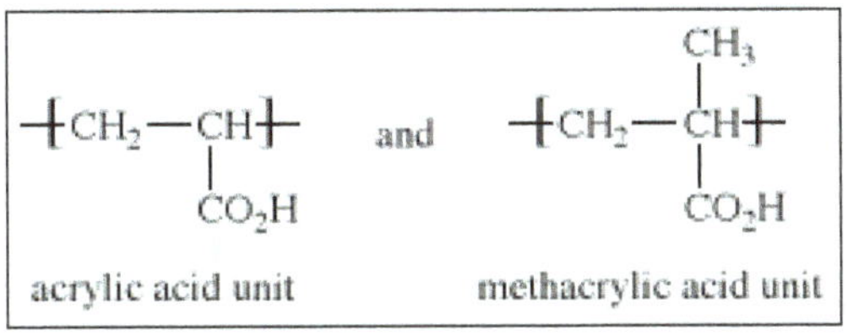

The acidic carboxyl (CO_2H) groups in these units are neutralized with bases to form highly polar ionic groups distributed along the polyethylene chains. These groups, drawn together by their electric charge, cluster together in "microdomains," stiffening and toughening the plastic without destroying its ability to be molded to permanent shapes. (Ionic polymers of this type are called ionomers.) The ethylene-acrylic acid and ethylene-methacrylic acid ionomers are transparent, semicrystalline, and impervious to moisture. They are employed in automotive parts, packaging film, footwear, surface coatings, and carpet backing. One prominent ethylene-methacrylic acid copolymer is Surlyn, which is made into hard, tough, abrasion-resistant golf-ball covers. Other important ethylene copolymers are the ethylene-propylene copolymers.

Polyvinyl Chloride

Polyvinyl Chloride (PVC or Vinyl) is an economical and versatile thermoplastic polymer widely used in building and construction industry to produce door and window profiles, pipes (drinking and wastewater), wire and cable insulation, medical devices etc. It is the world's third largest thermoplastic material by volume after polyethylene and polypropylene.

It is a white, brittle solid available in powder form or granules. Thanks to its versatile properties such as lightweight, durable, low cost and easy processability, PVC is now replacing traditional building materials like wood, metal, concrete, rubber, ceramics, etc. in several applications.

PVC Manufacturing Process

In 1872, E. Bauman exposed vinyl chloride sealed in a tube to sunlight and produced a white solid called PVC. By World War I, Germany was producing a number of flexible and rigid PVC products which were used as a replacement for corrosion-resistant metals.

Vinyl chloride monomer (VCM) is produced from the chlorination of ethylene and pyrolysis of the resulting ethylene dichloride (EDC) in a cracking unit. PVC (glass transition temperature: 70-80 °C) is produced by polymerization of vinyl chloride monomer (VCM).

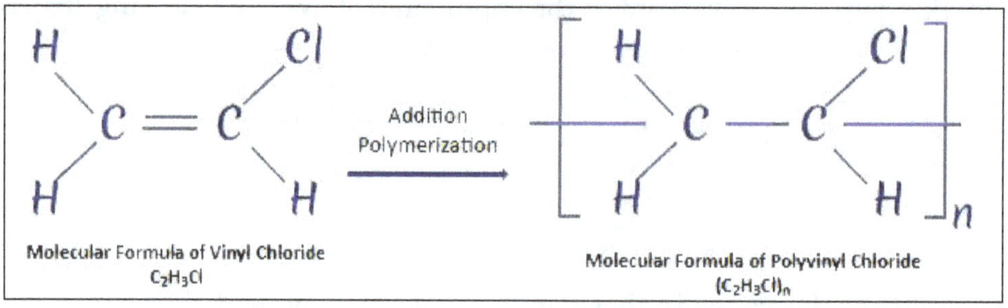

The popular methods used to manufacture PVC commercially are: Suspension PVC (S-PVC) Process and Bulk or Emulsion (E-PVC) Process. Suspension Polymerization accounts for 80% of PVC production worldwide.

Suspension PVC (S-PVC) Process

In pressure-tight reactor, the monomer is introduced with polymerization initiator and other additives. The content of the reaction vessel are mixed continuously to maintain suspension and ensure uniform particle size of PVC resin.

Typical suspension polymerized PVC has a mean particle size of 100-150 μm with a range of 50-250 μm.

S-PVC grades are formulated to meet an extensive range of requirements such as, high plasticizer absorption for flexible products, or high bulk density and good powder flow required for rigid extrusion

Bulk or Emulsion (E-PVC) Process

In this process, surfactants (soaps) are used to disperse the vinyl chloride monomer in water. The monomer is trapped inside soap micelles are protected by the soap and polymerization takes place using water soluble initiators.

The primary particles are solid, smooth surfaced spheres which are clustered into irregular shaped aggregates with a typical mean particle size of 40-50 μm with a range of 0.1-100 μm. E-PVC resins are used in a wide range of specialty applications such as coating, dipping or spreading.

Suspension PVC (S-PVC) Process	Bulk or Emulsion (E-PVC) Process
• Lower flexible PVC formula costs. • PVC particles obtained are mixed with plasticizers & can be extruded in pellets which are further used for processing via extrusion, calendering, injection molding. • Processing equipment is typically very expensive.	• Higher flexible PVC formula costs. • PVC powder obtained is mixed with plasticizers to produce a paste which is further used for coatings, dipping, spraying. • Processing Equipment may or may not be very expensive.

Chlorinated PVC (CPVC)

CPVC is manufactured by chlorination of PVC polymer thereby raising the chlorine content from 56% to around 66%.

Chlorination of PVC reduces the forces of attraction between the molecular chains. CPVC is also essentially amorphous. Both of these factors allow CPVC to be stretched more easily and to a greater extent than PVC above its Tg. Pipe (436), moldings (376) and sheet are formulated for high temperature use based on CPVC or blends of CPVC and PVC.

Properties of PVC Polymer

- Electrical Properties: PVC is a good insulation material, thanks to its good dielectric strength.
- Durability: PVC is resistant to weathering, chemical rotting, corrosion, shock and abrasion. It is therefore the preferred choice for many long-life and outdoor products.
- Flame Retardancy: Because of its high chlorine content, PVC products are self- extinguishing. Its oxidation index is ≥45. Antimony trioxide has been used extensively, usually in combination with phosphate ester plasticizers, giving excellent fire performance and mechanical properties.
- Cost/Performance Ratio: PVC has good physical as well as mechanical properties and henc provides excellent cost-performance advantages. It has long life span and need low maintenance.
- Mechanical Properties: PVC is abrasion-resistant, lightweight and tough.
- Chemical Resistance: PVC is resistant to all inorganic chemicals. It has very good resistance against diluted acids, diluted alkalis and aliphatic hydrocarbons. Attacked by ketones; some grades swollen or attacked by chlorinated and aromatic hydrocarbons, esters, some aromatic ethers and amines, and nitro-compounds.

PVC resin obtained from polymerization is extremely unstable due to low thermal stability & high melt viscosity. It needs to be modified before processing into finished products. Its properties can be enhanced/modified by adding several additives such as heat stabilizers, UV stabilizers, plasticizers, impact modifiers, fillers, flame retardants, pigments, etc. It shows a wide spectrum of properties ranging from the extremely rigid to very flexible.

Selection of these additives to enhance polymer properties is dependent on end application requirement. For example:

- Plasticizers (Phthalates, Adipates, Trimellitate, etc.) are used as softening agents to enhance rheological as well mechanical performance (toughness, strength) of vinyl products by raising the temperature. Factors that affect the selection of plasticizers for vinyl polymer are:
    - Polymer Compatibility.
    - Low Volatility.
    - Cost.

Flexible PVC Pipe.

- PVC has a very low thermal stability and stabilizers help prevent degradation of polymer during processing or exposure to light. When subjected to heat, vinyl compounds initiate a self-accelerating dehydrochlorination reaction and these stabilizers neutralize the HCl produced enhancing the life of polymer. Factors to be considered while selecting heat stabilizer are:
    - Technical requirements.
    - Regulatory Approval.
    - Cost.
- Fillers are added in PVC compounds for a variety of reasons. Today, a filler can be a true performance additive by delivering value in new and interesting ways at the lowest possible formulation cost. They help to:
    - Increase stiffness and strength.
    - Improve impact performance.
    - Add color, opacity and conductivity.
    - And more.

Calcium carbonate, titanium dioxide, calcined clay, glass, talc etc. are common types of fillers used in PVC.

- External lubricants are used to assist smooth passage of PVC melt through processing equipments. While internal lubricants reduce melt viscosity, prevent overheating and ensure good color of product.

- Other additives like processing aids, impact modifiers, are added to enhance mechanical as well as surface properties of PVC.

PVC Blend with other Thermoplastics

- PVC/Polyester Blends: These blends combine superior physical properties of polyesters with the excellent processing characteristics of PVC. Benefits include abrasion resistance, tensile properties and tear resistance.
- PVC/PU Blends: These blends offer increased abrasion and chemical resistance. Some TPUs are biocompatible and when blended with PVC results in valuable products for PVC industry.
- PVC/NBR Blends: Flexible PVC modified with NBR are melt processable yet possess good elasticity/recovery characteristics.
- PVC/polyolefin rubber alloys: They have potential utility in many applications where conventional flexible vinyl compounds do not meet certain end-use performance requirements.

Limitations of Polyvinyl Chloride

- Poor heat stability.
- Properties can change with time, due to plasticizer migration.
- Flexible PVC has lower chemical resistance than rigid PVC.
- Rigid PVC has low continuous service temperature of 50 °C.

Application of PVC Resin

Commercially, PVC is one of the most important thermoplastics in the world today. Rigid (unplasticized) PVC is one of the most widely used plastic materials. Main Applications of both types of PV (rigid and flexible) include:

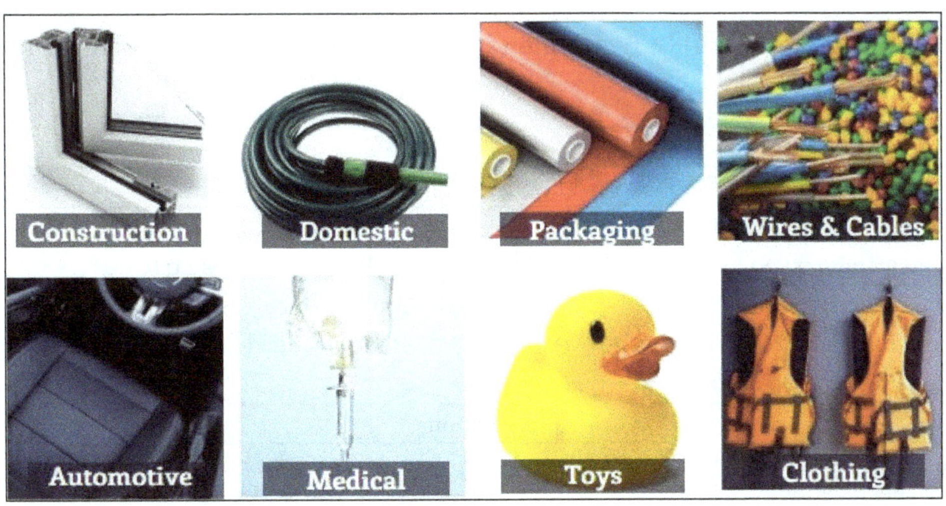

Application	Rigid PVC	Flexible PVC
Construction	Window Frames, Pipes, House Siding, Ports, Roofing	Waterproof Membranes, Cable Insulations, Roof Lining, Greenhouses.
Domestic	Curtain Rails, Drawer Sides, Laminates, Audio and Videotape Cases, Records	Flooring, Wall Coverings, Shower Curtains, Leather Cloth, Hosepipes.
Packaging	Bottles, Blister Packs, Transparent Packs and Punnets	Cling Film.
Transport	Car Seat Backs	Under Seal, Roof Linings, Leather Cloth Upholstery, Wiring Insulation, Window Seals, Decorative Trim.
Medical	-	Oxygen Tents, Bags And Tubing For Blood Transfusions, Drips and Dialysis Liquids.
Clothing	Safety Equipment	Waterproofs for Fishermen and Emergency Services, Life-Jackets, Shoes, Wellington Boots, Aprons and Baby Pants.
Electrical	Insulation pipes, Jacketing, Electricity Distribution Boxes, Switches, Transparent Distributor Box Housings, Plug Housings & Battery Terminals	Cable & Wire insulation, plugs, cable jackets, sockets, Sable Heads and Distributors.
Others	Floppy-Disk Covers, Credit Cards, Traffic Signs	Conveyor Belts, Inflatables, Sports Goods, Toys, Garden Hoses.

Processing of Vinyl Plastic

Some of the main processes include extrusion, calendering, injection molding, stretch blow molding, etc.

The intimate mixing of the PVC resin with its associated additives is necessary prior to converting into a thermoplastic melt. Thermal stabilization is required for processing rigid PVC, otherwise material may decompose during processing.

PVC is sensitive to the thermal history and the window of processing temperatures is quite small. Drying before processing is highly recommended, moisture rate should be lower than 0.3%.

Drying before processing is highly recommended for plasticized PVC, moisture rate should be lower than 0.3%.

Plasticized PVC	Rigid PVC
Injection Molding	
• Melt temperature: 170 and 210 °C • Mold temperatures: 20 to 60 °C • Mold shrinkage: 1 and 2.5% • Material Injection Pressure: Up to 150 MPa • Packing Pressure: Up to 100 MPa	• Melt temperature: 170 and 210 °C. • Mold temperatures:20 to 60 °C. • Mold shrinkage: 0.2 and 0.5%. • Recommended Screw with an L/D ratio of 15 to 18
Extrusion	
• Extrusion temperatures are 10-20°C below injection molding temperatures in order to avoid premature thermal degradation.	

PVC has largely been overlooked as being suitable for 3D printing, and the new developments are opening the way for PVC into the growing world of additive manufacturing. For example, Chemson Pacific Pty Ltd, a Vinyl Council of Australia member, demonstrated a world-first for 3DVinyl PVC material by 3D printing a giant flower vase using a pellet-fed 3D-printer.

PVC material can be bonded using different joining techniques to fabricate PVC into the finished article. All welding techniques involve the application or generation of heat to soften the material whilst pressure is applied simultaneously. Bonding techniques, involving adhesives, are also common.

Recyclability and Toxicity of PVC

Products made from PVC are 100% recyclable and can be identified as recycling code #3.

Adopting an appropriate recycling pathway is of both an economic value and an environmental benefit. Two key methods include:

- Mechanical Recycling: After mechanical separation, grinding, washing and treatment to eliminate impurities, it is reprocessed using various techniques (granulated or powder) and reused in the production.
- Feedstock Recycling: It involves (usually) thermal treatment of the PVC waste stream with recovery of hydrogen chloride that can then be returned to the PVC production process or used in other processes.

Recycled PVC can be used to produce packaging, film and sheet, loose-leaf binders, pipes, carpet backing, electrical boxes, cables and more. The industry is working with the regulatory authorities to ensure that recycling activities remain sustainable while complying with the regulatory regime.

Presence of chlorine content and use of additives such as plasticizers bought PVC under scrutiny for a number of years. Concerns have been raised at regular intervals, in several regions, regarding the possible negative impact of phthalates on the environment and human health.

Similarly, Europe has phased out use of lead-based stabilizers in vinyl compounds due to their classification as reprotoxic, harmful, dangerous for the environment and their presence (heavy metal) causing issues in waste management strategies.

Polypropylene

Polypropylene (PP) is a linear hydrocarbon polymer, expressed as CnH2n. PP, like polyethylene and polybutene (PB), is a polyolefin or saturated polymer. Polypropylene is one of those most versatile polymers available with applications, both as a plastic and as a fibre, in virtually all of the plastics end-use markets.

Properties

The properties of Polypropylene include:

- Semi-rigid.
- Translucent.
- Good chemical resistance.
- Tough.
- Good fatigue resistance.
- Integral hinge property.
- Good heat resistance.

PP does not present stress-cracking problems and offers excellent electrical and chemical resistance at higher temperatures. While the properties of PP are similar to those of Polyethylene, there are specific differences. These include a lower density, higher softening point (PP doesn't melt below 160 °C, Polyethylene, a more common plastic, will anneal at around 100 °C) and higher rigidity and hardness. Additives are applied to all commercially produced polypropylene resins to protect the polymer during processing and to enhance end-use performance.

		Homopol	Copolym
Density	/ kgm-3	905	905
Price / Tonne	/ £	680	620
Tensile Strength	/ Mpa	33	25
Tensile Modulus	/ Gpa	1.4	1.0
Elongation at Break	/ %	150	300
Hardness	/ Rockwell "R" Scale	90	80
Notched Izod Impact	/ kJm-1	0.07	0.1
Heat Distortion Temp (HDT)	@ 0.45 MPa / °C	105	100
Heat Distortion Temp (HDT)	@ 1.80 MPa / °C	65	60
Volume Resistivity	/ logÙm	19	19
Oxygen Index	/ %	17	17

Chemical Structure

Linear hydrocarbon polymer, little or no unsaturation. Similar to polyethylene in many properties, especially solution and electrical. However, the presence of the methyl group attached to every alternate backbone chain carbon atom can alter the properties in a number of ways:

- It can cause a slight stiffening of the chain - increasing the crystalline melting point (Tm).
- It can interfere with the molecular symmetry - depressing crystallinity and hence Tm.

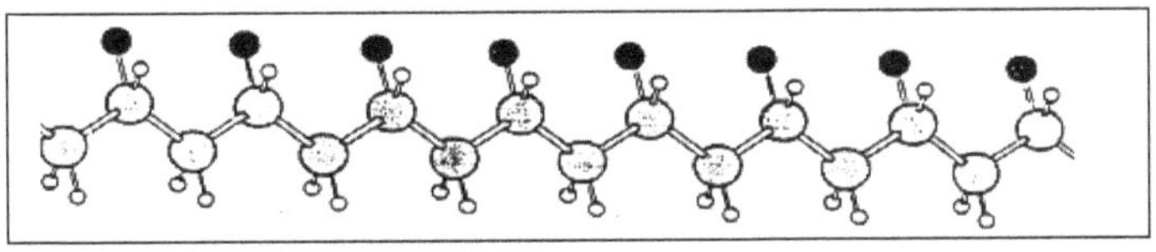

Isotactic

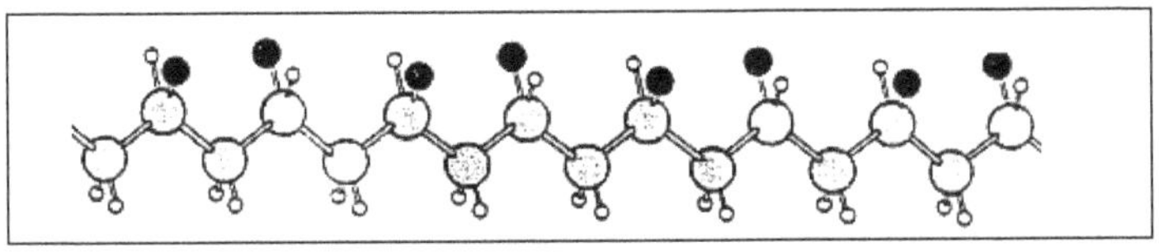

Syndiotactic

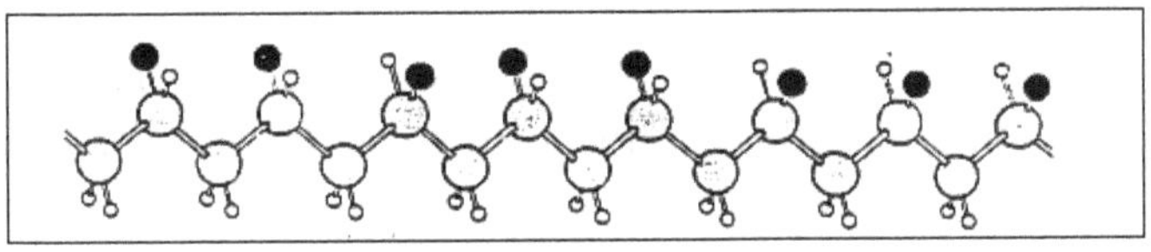

Atactic

In the case of very regular polymers, ISOTACTIC form, the net result is a melting point some 30 °C higher than that for HDPE.

The methyl group also has a chemical effect, i.e. tertiary carbon atom provides a site for oxidation - PP less stable than PE, also peroxides and radiation lead to chain scission rather than cross-linking.

Manufacturers quote "isotactic index" - amount insoluble in hot solvent - which indicates the percentage isotactic polymer in the grade. N.B. usually 90-95% isotactic. Glass transition around 0 °C presents problem of embrittlement and this can be altered by copolymerisation (with ethylene block copolymers) and/or blending with rubber, e.g EPM. Polypropylene does not crystallise so readily as polyethylene, it tends to supercool, and this becomes worse at higher molecular weights.

Manufacture

Production of polypropylene takes place by slurry, solution or gas phase process, in which the propylene monomer is subjected to heat and pressure in the presence of a catalyst system. Polymerisation is achieved at relatively low temperature and pressure and the product yielded is translucent, but readily coloured. Differences in catalyst and production conditions can be used to alter the properties of the plastic.

Propylene is obtained, along with ethylene, by cracking naphtha (crude oil light distillate). Ethylene, propylene, and higher alkenes are separated by low temperature fractional distillation.

From being a byproduct of ethylene in the early 1950's, propylene is now an important material in its own right.

Polypropylene is a major tonnage polymer with a growth rate higher than the norm for such thermoplastics. This growth rate is partially because of the versatility of polypropylene, and hence the wide range of application areas as indicated above. However, in the UK we have always used more polypropylene than in other European countries, e.g. West Germany. This is particularly the case in injection mouldings, which elsewhere might well be made from high density polyethylene.

Processing

		Homopolymer	Copolymer
Melt temperature	/ °C	210 - 290	210 - 290
Mould temperature	/ °C	20 - 60	20 - 60
Pre-drying	/ hours at °C	Not required	Not required
Typical mould shrinkage	/ %	1.5	2

Processing Characteristics

Polypropylene is a relatively easy materials to injection mould in spite of its semicrystalline nature. The absence of any real need for high molecular weight, from the mechanical properties view point, leads to low melt viscosity (easy flow). The pseudoplastic nature of polypropylene enhances this effect at high shear rates (fast filling rates).

Typically melt temperatures for injection moulding are between 200 and 250 °C. Though they can be as high as 280, or even 300 °C, for short periods of time. With Flame Retardant grades it recommended that 2200C is not exceeded. Mould filling rates are generally on the high side, to ensure good surface finish and strong mouldings free of weld lines and flow fronts. Adequate venting of the mould is essential to prevent burn marks.

The melt flow index gives a rough guide to melt flow behaviour, but as a result of the pseudoplastic nature of polypropylene (which is strongly dependant on the molecular weight distribution) this should not be taken too literally. Spiral flow mould data is sometimes provided by manufacturers as a practical means of assessing the flow behaviour, but this is no substitute for the fundamental melt rheological and thermal data which is now becoming more widely available.

Poly-caprolactone (PCL)

Polycaprolactone (PCL) is a polymer composed of hexanoate repeat units, included in the class of aliphatic polyesters. PCL has been thoroughly investigated for its peculiar mechanical properties, miscibility with a large range of other polymers, and biodegradability. The physical, thermal, and mechanical properties of PCL mainly depend upon its molecular weight and degree of crystallinity, which also contribute to its capability to degrade—by hydrolysis of its ester linkages—under physiological conditions. PCL is strongly hydrophobic, semicrystalline, highly soluble at room temperature, and easily processable due to the low melting temperature and exceptional blend compatibility,. Indeed, owing to its native biocompatibility and biodegradability, PCL has been extensively studied for the preparation of controlled drug delivery systems by several formulations of polymers or copolymers. Moreover, its permeability to a wide range of drugs enabled uniform drug distribution in the matrix, assuring a long-term release—up to several months—by a degradation mechanism. Owing to a markedly slow degradation rate, PCL has been largely used for the preparation of long-term implants and scaffolds able to reproduce the natural extracellular matrix, thus supporting the three-dimensional (3D) cell culture in tissue engineering and repair. Hence, PCL has also been certified as FDA approved (Food and Drug Administration, USA) and CE registered mark (European Community) for use in a large number of drug-delivery and medical devices, even though surprisingly few have been commercialized or widely translated to clinical studies.

More recently, PCL attracted interest for the design of green materials/biomaterials used for various applications. The superior rheological and viscoelastic properties over many of its aliphatic polyester counterparts render PCL easy to manufacture and manipulate into a large range of biodegradable devices. Moreover, PCL mechanical properties make it suitable for medical applications complementary to tissue engineering, such as, for example, wound dressing, contraceptive, and dentistry but also in non medical fields such as environment, packaging, and food. Much attention is currently put on the use of PCL in combination with biopolymers, owing to its interesting biofriendly features. These include tailorable degradation kinetics and mechanical properties, ease of shaping and manufacture enabling appropriate pore sizes conducive to tissue in-growth, and the controlled delivery of drugs contained within the matrix. In the latter case, functional groups could also be added to modify polymer chains to improve hydrophilic, adhesive, or biocompatible properties to induce favorable cell responses. In addition, PCL formulations have also been prepared by self-assembly to form micro-/nanomicellar hydrogel structures after copolymerization to make them amphiphilic, hence improving the encapsulation of bioactive molecules and drugs.

Synthesis

The first synthesis of PCL by thermal treatment of ε-caprolactone was reported by Van Natta and co-workers. Since the first approaches toward the realization of this biodegradable polyester, PCL is still mainly synthesized by ionic and metal catalyzed ring-opening polymerization (ROP) of the cyclic monomer ε-caprolactone, even if several works have been focused on the radical ring-opening polymerization (RROP) of 2-methylene-1, 3-dioxepane (MDO), using different conditions, and on the condensation of 6-hydroxycaproic acid.

As concerning the polycondensation of 6-hydroxycaproic acid, this synthetic route is only reported in a few papers. The most interesting approaches are based on the enzymatic synthesis, such as those based on the use of lipase. The use of Candida Antarctica Lipase B (CALB) immobilized on acrylic resin slowly gives rise to PCL with an average degree of polymerization (DPavg) of about 63 and a polydispersity of 1.6. Interesting results for the synthesis of PCL and the in situ production of PCL-based nanocomposites have been recently reported by using CALB immobilized on organo-modified nanoclay.

Another synthetic approach to PCL is the RROP of cyclic ketene acetals, proposed by Bailey and colleagues in the 1980s. Although studies available on this method are still limited, RROP is potentially very interesting for different factors. First, RROP allows the random combination of ester and vinyl units for the synthesis of several biodegradable poly(vinyl-co-ester)s using radical initiators such as azo-bis-isobutyronitrile or organic peroxides. During the RROP of cyclic ketene acetals in the presence of vinyl monomers, the reaction mechanism and the structure of the obtained copolymer strictly depend on the ring opening of the cyclic monomer over the 1, 2-vinyl addition, that is conditioned by the composition of the monomer mixture, the temperature, the ring size of the cyclic ketene acetal, and its substituents. For seven-member rings, such as MDO, the ring opening was almost quantitative even at low temperatures (50 °C).

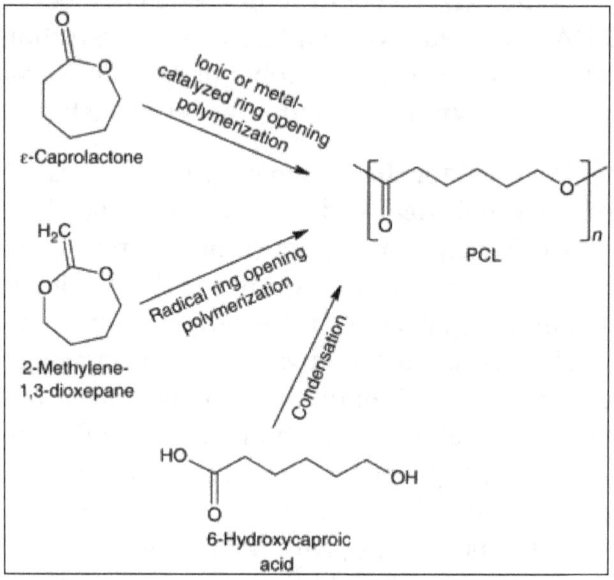

Synthetic routes to PCL by ionic or metal-catalyzed ROP of ε-caprolactone, by RROP of MDO and by condensation of 6-hydroxycaproic acid.

Agarwal and co-workers investigated the effects of the monomer ratio and the temperature on the kinetic and the mechanism of the copolymerization between MDO and vinylacetate, showing comparable reactivity of the two monomers. Other vinyl monomers copolymerized with MDO for the realization of functional biodegradable polyesters include styrene, methyl methacrylate, methyl acrylate, glycidyl methacrylate, dimethyl aminoethyl methacrylate, and propargyl acrylate. Recently, photoactivated cobalt-mediated radical copolymerization of vinylacetate and MDO has also been reported.

Nevertheless, it should be mentioned that several recent studies have been focused on reversible addition-fragmentation chain transfer (RAFT) polymerization (also reported as MADIX, macromolecular design via interchange of xanthates) to realize copolymers of MDO and vinyl or acrylate monomers with controlled functionalities, molecular weights, and polymer architectures. In addition, vinyl acetate derivatives, such as vinyl bromobutanoate, have been copolymerized with MDO using the RAFT/MADIX polymerization technique.

Another notable aspect of the RROP of cyclic ketene acetals, and in particular of MDO, is that the structure of PCL obtained with this method is different with respect to that obtained by ionic or metal catalyzed ROP of ε-caprolactone. A detailed investigation on PCL obtained by RROP showed 1, 4- and 1,7-H-transfer reactions leading to branched structures. The branching is a relevant phenomenon, because, depending on the experimental conditions, up to 20% branch density can be obtained. In some cases, the 1, 4-H-abstraction reactions can be avoided, bringing to PCL in which only 1, 7-H-abstraction reactions are present. The presence of these branching affects the crystallization of PCL, bringing to a highly amorphous polymer. This property has been used to realize blends of semicrystalline PCL obtained by ROP and amorphous PCL obtained by RROP, and these blends have shown a drastically increased compostability with respect to traditional PCL, even at low amounts of RROP–PCL.

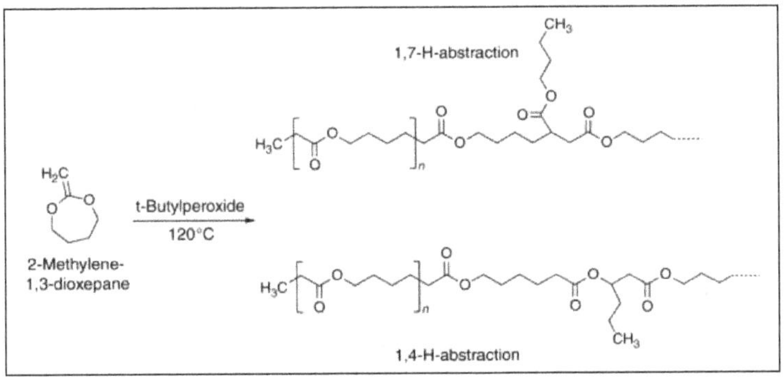

Branched structures in PCL obtained by RROP of MDO through 1,7- and 1,4-H-abstraction.

As stated, ROP of ε-caprolactone is the preferred route for the synthesis of PCL. The cyclic monomer is industrially obtained by the Baeyer–Villiger oxidation of cyclohexanone with hydrogen peroxide in the presence of different catalysts/initiators. The main mechanisms for the ROP of lactones (ionic and coordination–insertion), depending on the initiator used, are shown schematically in figure. They have been well summarized by Labet and Thielemans, that have already reported an exhaustive list of the experimental conditions used for the synthesis of PCL, including solvents, concentration ratios between monomers and catalysts/initiators, temperatures, reaction times, conversions, molecular weights, and polydispersity indexes of the realized polyester.

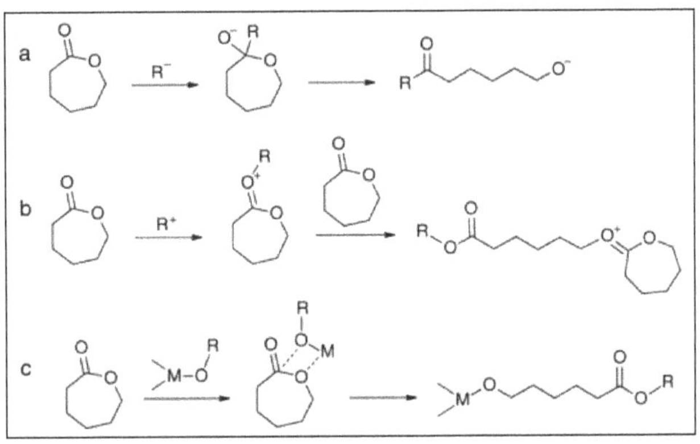

Mechanisms of the initiation steps of (a) anionic, (b) cationic, and (c) coordination-insertion ROP of ε-caprolactone.

During ROP, intramolecular (backbiting) and intermolecular transesterification side reactions must be considered, significant at the late stages of polymerization, inducing in general a loss in the control of the reaction and in particular an increased polydispersity.

As concerning ionic ROP, anionic and cationic ROP are reported. Anionic ROP is based on anionic species able to attack the carbonyl carbon of ε-caprolactone, inducing the ring opening and the formation of a growing alkoxide. Intramolecular transesterification is a relevant side reaction of this mechanism. An example of anionic ROP is found in the presence of alkali metal–based initiators.

Cationic ROP is based on the attack of a cation to ε-caprolactone with the formation of a positively charged species, which is then attacked by another monomer that induces the ring opening through a SN2 mechanism. An example of cationic ROP is found in the presence of Lewis acids.

The most common ROP used for the synthesis of PCL is based on the coordination–insertion mechanism, which proceeds through the coordination and the reaction of ε-caprolactone to a metal-based catalyst and the formation of a growing species linked to the metal atom by an alkoxide bond.

An extensive use of metal complexes is reported for the synthesis of PCL based on the coordination–insertion mechanism. These complexes allow to obtain very high molecular weights (up to 800,000 g mol−1) and low polydispersity, close to 1.1. In particular, tin(II) 2-ethylhexanoate and aluminum(III)isopropoxide are widely used catalysts.

More recently, the use of several transition metal catalysts (based on titanium and zinc) and rare earth metal catalysts has seen an even growing interest. Apart from the synthesis of the homopolymer PCL, research has been widely focused on the synthesis of several classes of copolymers, to tailor material properties toward specific aims. With this objective, copolymers of PCL have been investigated for a wide range of comonomers for biomedical applications including tissue engineering and drug delivery. In particular, these synthetic biodegradable polymers, using the approach described in a comprehensive review, can be classified in three main groups: polyesters, polymers containing both ester and other heteroatom-containing linkages in the main chains, and polymers with heteroatom-containing linkages other than ester linkages in the main chains. For instance, the synthesis of di- and tricomponent polyesters, including poly[(l-lactide)-co-(ε-caprolactone)] and poly[glycolide-co-(l-lactide)-co-(ε-caprolactone)], has been proposed to realize materials combining the elasticity of PCL and tailored degradation times.

Moreover, several research studies have been focused on PCL copolymers for the realization of materials with advanced molecular architectures, such as those presented in figure below.

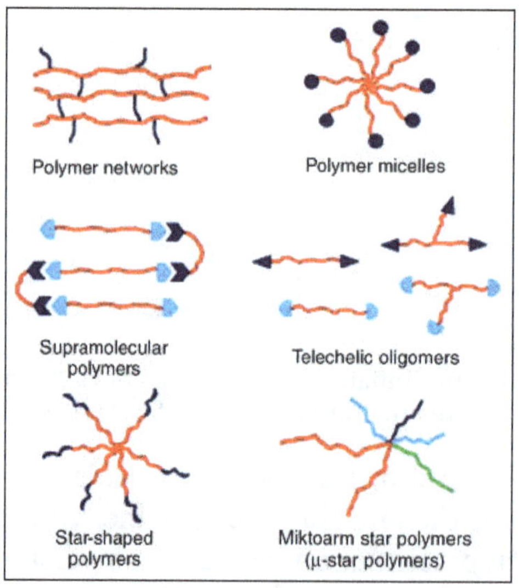

Representative advanced polymeric architectures employing PCL segments. PCL segments are represented by red polymer chains.

Polymer Processing

PCL Melting Processes

Extrusion is the most used technology in plastics industry since it is a continuous, very flexible process and allows high throughput. This technology has also been applied to the processing of PCL.

The extrusion process aims at melting the polymer to give a proper shape (film, plate, tube, and so on) and/or mixing with additives to obtain compounds or blends with improved mechanical or functional properties. It is usually based on one or two rotating screws that convey the polymer along a plasticating barrel through a die, specifically designed to take into account the polymer viscoelasticity. The die is responsible for the polymer forming to get the desired shape, which can be a finished or semifinished product. The extruder can be used alone or as a part of a more complex polymer treating plant, such as in the injection molding or foaming processes, or as a part of a multiextruder line.

One of the major extruder design parameters is the number of screws. In general, single-screw extruders are preferred to twin-screw types for their cost (roughly half the price of a twin-screw extruder) and their easier understanding and maintenance. Single-screw extruders are also preferred in the production of biodegradable polymers since they are characterized by a lower mixing intensity, which translates in lower shear stresses applied to the polymer and hence in lower chemical degradation. Twin-screw extruders are preferred for their higher performances such as greater and more steady feeding, higher throughput, higher mixing efficiency (but with higher stress transmission to the melted compound), and better dispersion of additives, better temperature control, and homogenization. The twin-screw extruder is also frequently used for foaming since it is able to quickly and more efficiently disperse the blowing agent, responsible for the formation of the porosity, and it can easily guarantee a strong and stable dynamic sealing before gas injection points.

The extruder has the capability to sequentially perform subprocesses along its length. These can be specifically performed by properly designing the screw profile, in particular recurring to segmented designs. Segmented screws are highly desirable because elements can be strung onto screw shafts to optimize subprocesses and can be changed accordingly to the specific processing needs and polymers, without the need to replace the entire screw. Both single- and twin-screw extruders can be used to process PCL, and their choice have to be related to the specific performance needs of the final product.

The main components of an extrusion line can be schematized in three conceptual zones: polymer feeding, plasticating, and feeding through the shaping die. Usually more complex designs are used to correctly perform the process, in particular if blends, compounds, or foams have to be produced. The extruder often presents the following zones:

- Materials feeding.
- Melting and compounding of additives.
- Gas injection and solubilization (in case foams have to be produced).
- Melt homogenization and cooling.
- Metering through the die.
- Die and postdie forming.

In the case of PCL extrusion, the melting, pumping, and forming subprocesses have to be performed in a process window that avoids polymer or additives degradation. This particular care has effects on a lot of parameters, such as the choice of the screws types and segmenting, formulation modifications, and operating conditions. To reduce the mechanical stresses on the polymer melt but at the same time to guarantee a correct dispersion of all the components, static mixers or open meshing elements can be used. The latter are preferred since the pressure rise is distributed over a larger length of barrel, and the heat transfer and thermal homogenization are obtained with lower energy consumption.

A homogeneous temperature of the melt at the die exit is very important since the viscoelastic properties of the extrudate are extremely sensitive to it. The production of finished or semifinished products needs a precise control of the exit temperature, which usually has to be lowered to increase the melt strength of the polymer. This task is particularly difficult to obtain if a single extruder is used since it conflicts with the plasticating function of the extruder and is typically addressed by using a long barrel. If a fine control of the processing conditions is needed and the higher cost can be managed, a second extruder operating in tandem with the plasticating one is the best solution. This solution is particularly effective since it allows to decouple the plasticating/mixing tasks from the shaping ones or further processing (eg, foaming), operated by the second extruder that takes care of the melt cooling to a temperature range where good quality products can be obtained.

If only melting and moderate mixing are needed, a single-screw extruder would be adequate. However, in most cases different components have to be intimately dispersed in the polymer and their agglomeration has to be carefully avoided. This is generally better addressed by using a twin-screw extruder, which may deagglomerate and disperse the nucleating agents, colorants, as well as fillers in a controlled way that ensures homogeneous properties throughout the extrudate. Furthermore, if blending of two polymers is part of the compounding process, the twin-screw design is quite mandatory.

PCL Blends

PCL has been widely blended with other polymers to increase its thermal, viscoelastic, or mechanical properties. Moreover, PCL has been used as a minor phase in binary or ternary systems to increase their functional properties. Among all biodegradable blends, PCL/PLA has been one of the most investigated. The main reason for such system is the possibility of an increase in the mechanical and thermal properties of PCL, which are weak for any industrial need that requires just a moderate tolerance to temperatures above room temperature. Blends can be obtained either by physical blending or by reactive extrusion. Since PCL and PLA are not miscible, the physical blend still presents two distinct glass transition temperatures. To overcome such a problem, they have to be compatibilized to take advantage of PLA peculiarities. Reactive compatibilization has been investigated by Wang and co-workers that prepared PLA/PCL blends by means of three coupling agents, aimed at exploiting a transesterification reaction to increase their compatibility. They succeeded in producing compatibilized blends, with higher elastic modulus but lower strength with respect to physical blends of the same composition. The main drawback was the increased degradation rate of the reactively compatibilized blends that was much higher than that of pure PLA and PCL. On the contrary, the degradation rate of physical blends is between those of pure PLA and PCL.

Blends of PCL and starch and its derivatives have been investigated by Koenig and Huang. They found that some degree of cross-linking was needed to improve the thermal properties of the blends. In not compatibilized systems, two phases were clearly detectable with both optical microscopy and thermal analysis. The phase morphology was typical of particle-reinforced systems, where starch particles were dispersed in the PCL matrix, and the advantages were an increase of the elastic modulus (+ 50% with respect to PCL) at the expenses of a reduction of the tensile strength (-15% with respect to the neat polymer). A reactively extruded starch–PCL nanocomposite blend was prepared by Kalambur and Rizvi, who found that the viscosities of nanocomposite blends were significantly lower than that of 100% PCL and nonreactive starch–PCL composites synthesized from simple extrusion mixing. Thermoplasticized starch (TPS) and PCL blends were prepared to overcome some weakness of pure TPS, such as viscosity and thermal degradability. The presence of PCL can affect significantly the rheological behavior of the blends, as well as it can improve the thermal stability of the blend.

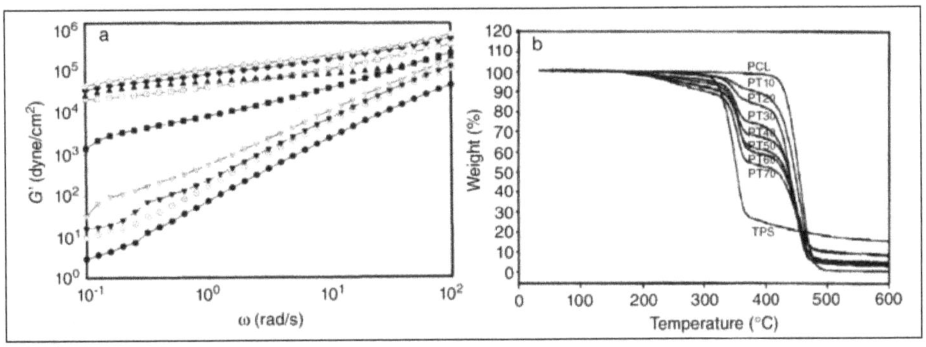

(a) Storage modulus versus frequency plot for PCL/TPS blends at 170 °C (PCL content decreases going from the lower curve to the upper one) and (b) TGA curves of PCL/TPS blends (PCL content decreases going from upper to lower curve).

Even at low PCL content, for example, 10 wt%, different mechanical (resilience) as well as functional (moisture sensitivity and shrinkage) properties are improved but they are significantly conditioned by the phase morphology. PEO/PCL blends were prepared for oral drug delivery applications. In this system, PCL was used to reduce the drug release rate profile. the performance of the blend was dependent on the processing conditions, and in particular on the shear stresses and temperatures applied during the mixing process. In fact, higher screw speeds were observed to result in slightly lower matrix melt viscosity when compared with matrices compounded using lower screw speeds. the immiscibility of PCL/PEO blends and found that the crystallization process of PCL was not affected by the PEO amount.

Blends of PCL and proteins have also been investigated. Proteins extracted from plants attract special attention in the production of biodegradable polymers. Proteins such as zein (corn), gluten (wheat), soy protein isolate (SPI), and peanut protein have been used. SPI has been widely investigated. Owing to the very difference affinity of proteins and PCL, a compatibilizer must be used, such as polyvinyllactam, methylene diphenyl diisocyanate, or modified cornstarch. The use of the compatibilizer leads to the increase of the elastic modulus and strength but lowers the plasticity (lower strain to failure). An effect of the addition of PCL is the reduction of water absorption and the increase of stability under ambient conditions relative to the plastics made from soy protein alone. John and colleagues modified PCL to incorporate a functional group that could interact with the functional groups on the gluten protein. Their results showed that a small amount of anhydride-modified PCL in the blend improved the physical properties of these blends over those of simple mixtures of wheat gluten and PCL. The presence of different phases and the use of high amount of proteins can result in a narrowed window of processability and decreased elongational strength.

Ternary systems, such as PCL–starch–low density polyethylene (LDPE), have also been prepared by Matzinos and others. The ternary blend was prepared by means of the extrusion process, which induced the thermoplasticization of the starch, and formed in film and molded parts. The mechanical performance was found to depend not only on the composition but also on the generated morphology. In films, the fine dispersion of PCL in the polyethylene/starch matrix resulted in increased mechanical properties, whereas in injection moulded specimens there was a decrease in properties due to phase coalescence. The ternary blend based on PLA/PCL/TPS was prepared by Sarazin and co-workers, who produced the blends by using a one-step extrusion process with the aim of exploiting PCL to increase the plasticity of PLA. The morphology and quantitative image analysis of the blends exhibited a three-phase morphology, with a fine dispersion of PCL "particles." The thermomechanical analysis clearly showed that the temperature of the tan δ peak of PLA is independent of TPS blend composition and that the addition of PCL in the ternary blend has little influence on the blend thermal transitions. The coupling of PLA with PCL and starch resulted in an increase in ductility (elongation at break up to 55% from 5% for the pure PLA) and of notched Izod impact energy, clearly indicating a synergistic effect that exceeds the results obtained for any of the binary pairs.

To improve the interface interactions between PCL and other polymers, different routes have been investigated such as polymer/polymer inclusion compounds, such in the case of PLA/PCL blends, or grafted copolymers. The managing of the compatibility between the two or more phases in a blend can result in the tailoring of additional characteristics such as rate of biodegradation and mass transport of low molecular weight species (water vapor, CO_2) other than thermal transitions and mechanical behavior. PCL grafted onto starch through introduction of urethane linkages was very effective in increasing the mechanical properties with respect to not compatibilized blends. Furthermore, the blend did not show the typical two melting temperatures (T_m) of phase-separated systems because a single T_m after PCL was compatibilized. Also the crystallization temperature can be influenced and its depression can be controlled through the amount of compatibilizer added to the blend.

Processing via PCL Solution

PCL-based Porous Materials

In the past years, several techniques have been used to manipulate PCL in the form of porous matrices. Conventional fabrication techniques such as salt leaching, gas foaming, phase separation, and freeze-drying have been mostly used, despite they do not offer a precise control of internal scaffold architecture and their properties (i.e, mechanical ones) for the fabrication of complex architectures. The gas foaming is a processing technique that allows the fabrication of biodegradable foams with porous architecture suitable for green applications in different fields (i.e, biomedical, environment, energy). The main advantage of this technique is related to unique chance of forming the porous network avoiding the use of organic solvents that may be harmful in biological or biosafe environment. During the gas-foaming process, the polymer is saturated with a gas or supercritical fluid—usually CO_2, N_2, or their mixture—under constant processing conditions (i.e, temperature, pressure). When the solubilization of the blowing agent into the polymer is completed, the polymer/blowing agent solution is brought to the supersaturated state either by increasing temperature (i.e, temperature-induced phase separation) or by reducing pressure (i.e, pressure-induced phase separation), with the effect of inducing the nucleation and growth of gas bubbles into the polymeric matrix. These mechanisms play a key role in controlling—by an accurate setting of processing parameters—the final foam morphology. For instance, Xu and colleagues prepared biodegradable PCL foams in the presence of supercritical CO_2 gas. They reported that lower depressurization rate and saturation temperature led to increased foam bulk density.

Alternatively, PCL porous matrices have largely been fabricated via thermal induced phase separation (TIPS), a complex process depending on the thermodynamics and the kinetics of the polymeric solution during cooling. The basic idea provides the cooling of a polymer solution to induce a phase separation in two phases, respectively, one polymer-rich phase and one polymer lean phase. Later, the removal of the solvent within the polymer lean phase by solvent evaporation, sublimation, or solvent/not solvent exchange allows reaching an open pore network whereas the polymer that composes the polymer-rich phase solidifies in the final structure. Typically, a liquid–liquid phase separation occurs when the imposed temperature is higher than the solvent crystallization one or freezing point whereas a solid–liquid phase separation takes place when the solvent crystallization temperature overcomes the cooling one. Pore characteristics (i.e, microtubules diameter, shape, and orientation) and structural defects may be controlled by an accurate definition of thermodynamic parameters and mold manufacturing. In the case of PCL matrices, several studies investigated the formation of anisotropic regions due to the application of local "not intentional" temperature gradients, which promote the growth of microsized channels—from the surface to the inner regions—orthogonally oriented to the solidification front. This peculiar architecture can be strictly controlled by the design of custom-made molds with insulated walls able to minimize the formation of transversal gradients and promoting a preferential heat conduction in the longitudinal direction to obtain oriented microtubular porosity able to confer peculiar anisotropic properties (i.e, mechanical response) to the polymeric structure.

Salt leaching is based on the addition of porogen or salt crystals (eg, sodium chloride) to a polymer solution arranged into a stable mold. This slurry (i.e, polymer plus porogen) is subsequently hardened by solvent removal after phase inversion, evaporation, or separation, and only then salt is removed via dissolution in water or alcohol thus forming a porous structure once all the salt leaches out. Pore size and pore size distribution can be controlled by the size and size distribution of the porogen used, whereas the porosity can be controlled by varying the amount of the salt particles. These parameters have relevant effects on the mechanical response, consistently with changes in morphological features. However, pore shape and interconnections are generally not controllable so that particle leaching is generally combined with other scaffold fabrication techniques, to create porous matrices fully percolative to fluids and small molecules. Besides, similar drawbacks have also been recognized in other processing techniques. For instance, porous PCL matrices fabricated via gas foaming generally show a remarkable lack in pore interconnection, strictly depending upon the gas expansion mechanism during foam formation and the inherent high melt strength of PCL. In the case of PCL matrices obtained by TIPS, pore sizes are generally smaller than 100 μm and are usually unsuitable to accommodate different cells which have to stretch and adequately grow, impairing the cell proliferation within the scaffold.

Hence, other technological approaches (i.e, rapid prototying, electrofluidodynamics) have been more recently implemented to design micro- and nanostructured platforms with fully percolative structure by using biodegradable polyesters such as PCL in solution, to overcome the main limitations of conventional techniques. Rapid prototyping techniques supported by computer-aided design (CAD) modeling have largely demonstrated to be feasible and consistent for a finest control of the PCL scaffold architecture at both micro- and macrolevels, but still present some relevant limitations in terms of resolution. Among them, stereolithography allows photopolymerizing a liquid photo-cross-linkable resin, such as PCL monomers in a chemically modified solution, into designed 3D structures with the highest accuracy and precision by successive deposition of thin layers, photoirradiated by ultraviolet or visible light according to a sliced CAD model. In this case, resolution of each layer is dependent on the resolution of the elevator layer and the spot size of the laser.

However, owing to the application of an additional curing step to improve the model's property, the final resolution may be compromised by the shrinkage that typically occurs in this postprocessing step. Alternative approaches based on the physical interaction of PCL solution with electrostatic forces (i.e, electrospinning) allows prefabricating controlled textured matrices with controlled strut sizes; however, several limitations arise for pore sizes over the micrometer scale. Electrospinning is based on electrohydrodynamic principles relying on an electrified viscous fluid jet being drawn through the air toward a collector at a different electric potential. PCL fibers fabricated via electrospinning generally present average diameters typically ranging from 0.5 to 2 μm. However, the chaotic nature of fiber deposition commonly results in tightly packed nonwoven meshes with pore sizes too small to assure an efficient penetration of micrometric or submicrometric objects (i.e, cells and molecular complexes). To increase the percolation or invasiveness index of these structures, PCL may be processed in the form of melt solution that differently interacts with electrostatic forces so producing much larger sized filaments up to 250 μm in size. More recently, PCL scaffolds by covalent attachment of a hydrogel component to modified methacrylated PCL segments were fabricated through electrofluidodynamic writing processes combining unique benefits of 3D printing technique, electrospinning, and direct writing mode.

PCL-based Micro- and Nanoparticles

Various preparation methods have been used in the past years for the fabrication of PCL microspheres and the encapsulation of different drugs. They include phase separation, emulsion evaporation, solvent extraction, spray drying, and melt encapsulation. In all cases, it is possible to obtain spherical microspheres with average sizes ranging from 50 μm to 2 mm. Drugs or bioactive molecules can be entrapped or encapsulated with different efficiency as a function of the used processing method. The drug may be released by matrix erosion and/or by diffusion trough the matrix or the shell forming a microsphere reservoir. Their large surface-to-volume ratio and rigidity in shape allows controlling drug release kinetics as a function of the peculiar properties of the polymeric matrix. More in general, PCL microspheres with modulated release kinetics can be obtained either by selection of appropriate preparation method, processing condition, and/or change in polymer composition leading to a altered degradation behavior. As a consequence, PCL microspheres have been considered for several applications, other than in controlled drug delivery such as for drug targeting upon derivatization or surface modification.

As first attempt, PCL microparticles have been prepared by a polymerization of colloidal monomers dispersed in a liquid with opposite solubility. More commonly, microparticles can be processed from dispersed droplets of monomeric solutions by various methods including emulsion, suspension, and dispersion techniques. The water (W)/oil (O) solvent evaporation method is the simplest solution, but it is suitable for lipophilic drugs only; whereas W/O/W is useful in encapsulating both oil and water soluble drugs. Alternatively, oil in oil (O/O) and water in oil in oil (W/O/O) methods have been largely used for the encapsulation of any water-soluble drug—either alone or in combination with lipophilic drugs. By these techniques, it is possible to fabricate uniform spheres from the macro to submicrometric scale (0.1–100 μm). In the case of micrometric sizes, spherical droplets can be formed by oil-soluble organic solution of PCL dispersed in aqueous media—oil in water (O/W)—or by water-soluble monomers dissolved in water dispersed in an organic medium—water in oil (W/O). To reduce particle sizes down to 1 μm or below, polymer beads can be prepared by dispersion polymerization into a range of 0.5–10 μm, or by suspension polymerization down to 50–500 nm, a size only detectable via visible light diffraction. In these cases, reagents including monomers, initiators, and stabilizers are dissolved in an organic medium and, since the initiator is soluble inside the monomer polymerization takes place inside the monomer droplet. Polymer beads, insoluble in the organic solvent, begin to precipitate, and the stabilizer prevents bead flocculation. Size and particle quantity have to be determined as a function of the relative phase ratios and the speed of mechanical stirring. To improve the encapsulation of bioactive molecules, a double emulsion strategy can be adopted to preliminarily dissolve biomolecules into water prior to disperse them into an organic solvent (usually dichloromethane, DCM, for PCL). Poly(vinyl alcohol) or other surfactants may be used to stabilize this second emulsion. In this case, microspheres are formed as DCM evaporates and the polymer hardens, thus trapping the encapsulated drug.

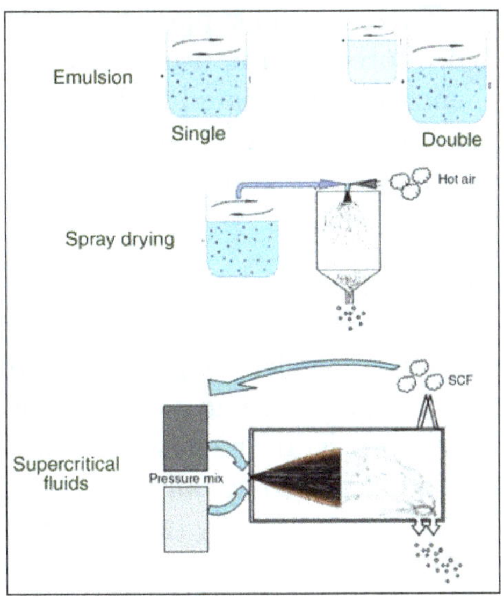

Schematic of the main processing techniques used to fabricate PCL micro- and nanoparticles for drug delivery applications.

To avoid the use of organic solvent, alternative strategies including hot melt technique and spray drying have been adapted, case by case, to the use of thermostable drugs. Spray drying techniques have been commonly used for producing microspheres from linear polymers. In place of spray drying, freeze-drying is proved to be more advantageous in terms of decreased particle size and increased burst release.

Only recently, novel processes based on supercritical fluids such as rapid expansion, antisolvent precipitation, and solvent-enhanced dispersion are emerging as innovative approaches to encapsulate a wide range of drugs, minimizing any processing traces in the final formulation for the fabrication of a fully recycling product.

Applications

Scaffolds for Tissue Engineering

Tissue engineering strategy is based on the design of porous scaffolds with peculiar biodegradability/bioresorbability of pores, to be used as 3D template for supporting cell attachment and subsequent tissue formation, both *in vitro* and *in vivo*. 3D scaffolds have to present suitable features for the creation of a cell friendly microenvironment able to exert mechanical support, physical and biochemical stimuli for optimal cell growth and function. For this purpose, it is mandatory an accurate selection of constituent polymers in terms of degradation, mechanical properties, workability by commonly used scaffold fabrication methodologies, and intrinsic biocompatibility with cells under different culture conditions. For instance, a bioinspired scaffold has to degrade up to *in vivo* resorb at a predefined rate so that the 3D space occupied by the initial scaffold can progressively be replaced by the regenerated host tissue, minimizing the inflammatory response of the substrate. In this context, some aliphatic polyesters such PCL are very attractive in tissue engineering because of their good biocompatibility and processability. In particular, PCL homopolymer degrades extremely slowly by hydrolysis, showing low values of water absorption and weight loss after 110 weeks in an aqueous medium simulating body fluids. This behavior endows the PCL matrix with sufficient mechanical properties to sustain reduced loads such as the hydrostatic pressure of biological fluids inside the structure, offering a valid support function until its complete degradation, when tissue will be totally formed. Moreover, the use of a long-term degradable material such as PCL avoids stress-shielding phenomena associated with the use of rigid implants. Indeed, the progressive degradation of the implant may trigger a gradual stimulation of healing mechanisms —that is, remodeling in the case of bone—by a gradual transfer of physiological loads, thus minimizing stress-shielding atrophy.

Besides, it is mandatory to impart peculiar functionalities able to imitate, or at least, to regenerate the native physiochemical environment of tissues. This is possible by imparting specific morphological and biochemical cues to PCL materials by providing an accurate manipulation of pore architecture at the micro and nanoscale, and/or chemical modifications to incorporate bioactive phases (i.e, bioceramic) or molecular signals. In the past years, several studies have been focused on the optimization of topological features such as full pore interconnectivity, tailored pore size, and shape that can actively support cell functions by regulating the interaction between the cells and the diffusion of nutrients and metabolic wastes across the whole 3D construct. It is well known that an optimum pore size range may be selected as a function of cells or tissues to be regenerated such as 5–15 μm for fibroblast in-growth, 20–125 μm for skin regeneration, 100–300 μm for bladder smooth muscle cell adhesion- and in-growth, 100–400 μm for bone regeneration.

Hence, processing techniques have to be adequately selected to confer tunable porosity as well as mechanical properties able to carry out the most appropriate biological response to PCL scaffold. For instance, temperature-driven processing routes, such as TIPS and melt co-continuous polymer blending, may be adopted, alone or in combination with conventional techniques (i.e, particulate leaching) to fabricate multiscale porous scaffolds with peculiar structure/property combinations. This mainly depends upon the peculiar chemical/physical properties of PCL—that is, low melting point, solubility in several organic solvents—which allows for their easy processability under controlled thermal conditions.

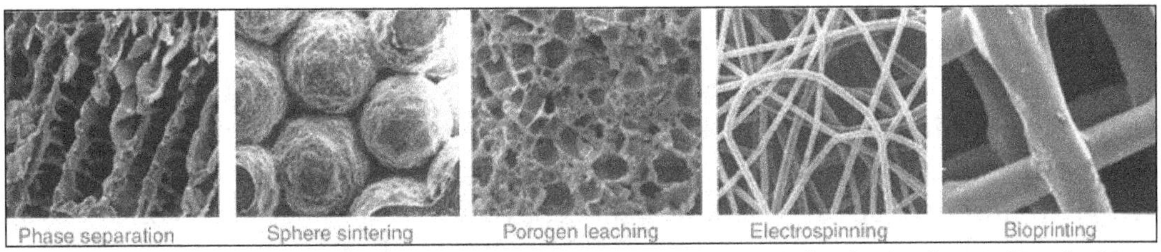

Porous scaffolds made of PCL fabricated by different processing technologies from PCL solution.

For instance, in the case of PCL scaffolds fabricated by TIPS, a judicious setting of solvents, cooling modalities (i.e, temperature and cooling rate) enables guiding the thermodynamic phenomena associated with the formation of dispersed phases and thus giving the chance to obtain either random or oriented pore architectures to mimic the anisotropic structure of natural tissues. Alternatively, a temperature-controlled extrusion of binary blends including polymers with a similar melting point (i.e, PCL and PEO) may allow forming co-continuous 3D microstructures as a consequence of the interpenetration of two different melt phases able to reproduce fully interconnected anisotropic pore networks with improvements in the mechanical properties and fluid permeability with respect to conventional scaffolds processed by polymer solutions (i.e, salt leaching). This result is really promising to realize structurally organized platforms able to guide the regeneration process of hierarchically organized tissues (i.e, tendons, muscles, ligaments, and nerves).

Biodegradable Carriers for Drug Delivery

Biodegradable polymers currently represent one of the most interesting opportunity to design drug delivery systems due to their relevant advantages, such as the easy removal of polymer metabolites out of the body by innate metabolic processes, and the unique chance to process them in the form of micro- and nanoparticulate systems to finely control site-specific and targeted drug delivery. To date, drug administration via degradable drug delivery systems is advantageous, giving the opportunity to design polymer microcarriers easy to be ingested or injected. For instance, they can be tailored for organ-targeted release, by improving the biological activity, controlling the drug release rate, and decreasing the administration frequency. This is possible because the polymer matrix can biodegrade within a suitable period, which is compatible with the drug release rate, so properly dosing active principles in the site of interest, improving the therapeutic efficiency and efficacy, but minimizing any side effect during pharmacological treatments.

Hence, an extensive research has been carried out to explore the suitability of biodegradable polymers not only to develop bioresorbable devices (i.e, surgical sutures) but also drug delivery systems. Among the different classes of biodegradable polymers, PCL has been extensively studied in several formulations not only for their recognized biocompatibility and biodegradable nature but mainly for its large variability of degradation properties, which allows fruitfully using it for different pharmaceutical dosage forms. Its compatibility with a wide range of drugs enables uniform drug distribution in the matrix, whereas its long-term degradation facilitates drug release up to several months. The advantages of PCL include its high permeability to small drug molecules, and its negligible tendency to generate an acidic environment during degradation as compared to other polyesters such as PLA and polyglycolic acid (PGAs).

The degradation of PCL homopolymer is very slow compared to other polyesters, making it more suitable for long-term delivery systems extending to a period of more than 1 year, despite it can be increased/decreased by proper physical or chemical modifications. Meanwhile, the high permeability to many drugs supports its ability to be fully excreted from the body once bioresorbed. Drug release rates from PCL substrates may depend upon a large pattern of parameters as function of the specific formulation and preparation methods, such as PCL content, size, and percentage of relative drug fractions used. Owing to the high workability, PCL may be easily blended with other polymers to improve stress, crack resistance, dyeability, and to better control drug release rates. However, blending strategies generally lead to altered physical properties and biodegradation with relevant effects on mechanical properties, so that this approach is generally preferred for the preparation of platforms for tissue engineering, such as porous scaffolds, fibers, or films. In the past years, PCL has been successfully combined with natural polymers, such as starch, gelatin, collagen, or chitosan, as well as synthetic ones including polyethylene glycol, polyurethane, and polyvinyl alcohol, to satisfy the main requirements in terms of biophysical/chemical properties for drug-dispensing formulations currently used in drug delivery.

Several studies have underlined some relevant constrains about the release mechanisms of therapeutic compounds, which may be significantly hindered by the poor water solubility of PCL. However, recent advances in drug formulations based on the use on innovative colloidal vectors currently concur to improve the drug delivery system performance, by providing a more efficient dispersion of drugs in PCL matrices by the use of solubilizing agents. For example, the formation of block copolymer micelles—a hydrophobic core sterically stabilized by a hydrophilic corona—may allow for the increase of the solubility of hydrophobic molecules by the peculiar polar interaction with local groups. In this case, chemical, physical, or electrostatic interactions allow variously entrapping drugs as a function of the physicochemical properties of macromolecules in the inner core, thus working as reservoir systems. Within the past decade, similar approaches have been used especially to develop controlled delivery systems for peptides and proteins.

Other Medical Devices

PCL is also suitable for the fabrication of innovative devices that are only indirectly interfaced with natural tissues, thus requiring an intermediate level of biocompatibility.

For instance, several studies have been performed in the past 40 years to identify the best material to be used for suturation. To date, sutures made from aliphatic polyesters including PGA (Dexon™), polylactic-polyglycolic acid (PLGA 10/90 - Vicryl®), and polydioxanone (PDS) have been broadly commercialized with different benefits as consequence of their specific biocompatible features. Indeed, the main problem of suture materials concerns their inflammatory response. It has been proved that Dexon™ and Vicryl® are more invasive on the activity of different kinds of cells with respect to PDS. Only recently, PCL has been investigated for the fabrication of suture filaments. Despite their limited applicability due to their degradation *in vitro* via the bioerosion mechanism, different copolymers have been investigated—that is, block copolymers of PCL with glycolide commercialized as Monacryl®—to obtain monofilament sutures with reduced stiffness compared with pure polyglycolide, commercialized by Ethicon, Inc.

PCL is also one of most used resorbable polymers for wound-healing applications. In most cases, PCL is fabricated in the form of microcarriers for *in vitro* and *in vivo* subdermal delivery of bioactive drugs (i.e, l-methadone) or ultrathin film for the release of chemical antiseptic chlorohexidine by dressing cutaneous wounds.

In the past two decades, relevant studies have been developed to design biodegradable matrix implants for controlled release of contraceptives to circumvent the need for device retrieval surgery. PCL is a highly desirable candidate for this role owing to its slow degradation, biocompatibility, and FDA approval. Dhanaraju and co-workers have prepared and characterized PCL microspheres as injectable systems for a controlled delivery of contraceptive steroids, directly into the implant site. Sun and others have developed a 2-year contraceptive device comprising PCL/Pluronic F68 compounds filled with levonorgestral powder, which was approved by the FDA to conduct phase II human clinical trials in China. Preclinical studies using rats and dogs demonstrated good release kinetics of levonorgestral from this device, with no adverse effects. After implant retrieval after 2 years, the implant was physically stable with an associated drop in the molecular weight of the polymer from 66,000 to 15,000 Da.

Other studies suggested the use of PCL for orthopedic applications. They mainly involve the design of composite fixation systems by reinforcing PCL with glass fibers. The main advantage of the use of PCL is related to the lower stress shielding with respect to metals (i.e, stainless steel), traditionally used for prosthetic devices. However, mechanical strength of PCL may be not sufficient for load-bearing applications and their use is mainly oriented to design resorbable composite implants for craniofacial repairing where more resilient materials are required. Furthermore, PCL has been reinforced with several different fibers including knitted polymer mesh or inorganic glass fibers.

Biodegradable Films for Packaging

Packaging applications are one of the most intensely studied applications for PCL-based materials with comparable functionalities to those of traditional oil-based plastic packaging. Despite current higher costs compared with the traditional plastic counterparts, many have found increasing commercial applications in packaging, in particular in the food industry.

PCL has preferably been used in films in blend form or coupled with other biodegradable polymeric films through compatibilizing layers due to the low elastic modulus and very fast (for packaging use) degradation. For example, the use of PCL and thermoplasticized zein (TPZ) blends has proved to have excellent adhesive capability in laminated PCL and TPZ-packaging structures as shown in figure. These laminated structures were found to be suitable for high pressure pasteurization treatments thanks to the combination of specific performances allowed by each layer of the film. Tests performed on multilayer structures confirmed the possibility to industrially use such packaging structure. A sterilizing treatment, obtained by means of a high pressure apparatus up to 700 MPa, did not promote any detectable change of oxygen and water vapor barrier properties of film. In figure, the appearance of pouches made of zein–PCL multilayer structures containing food before and after a high pressure pasteurization are reported and no visible degradation of the film can be observed.

Other examples of PCL-based films for food packaging are reported by Swapna Joseph and colleagues. The films were casted and dried at 55°C. The presence of PCL in chitosan resulted in an increase in elongation by 20.56% as compared to the pure chitosan film. PCL and chitosan showed to be miscible, and a co-continuous phase was detected in the 80:20 chitosan/PCL blend.

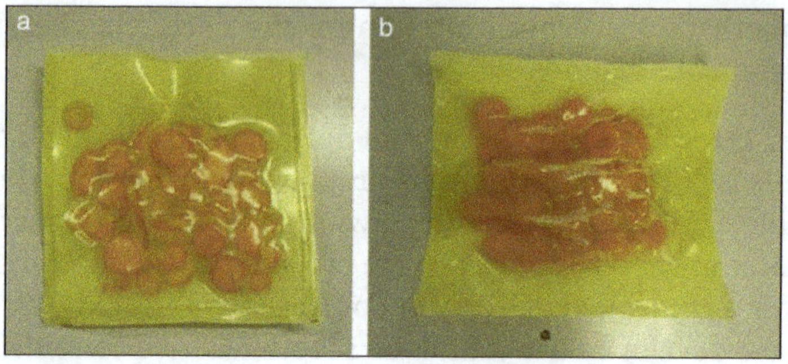

Zein–PCL package containing solid carrots (a) before high pressure treatment and (b) after high pressure pasteurization at 700 MPa.

On the other side, blending has proved to be an effective approach to improve the mechanical response of the film as a result of the use of PLA in the blend. The compatibility of antimicrobials, such as for the cyclodextrin-encapsulated type (allyl isothiocyanate), with PLA/PCL copolymer films was demonstrated and proved to be a viable solution to control the development of fungi on packaged cheeses, without incurring in substance migration issues since those species are not contaminant for food.

The development of an active packaging based on biodegradable matrices is a promising route to increase the use of biodegradable polymers in industrial packaging. Some studies investigated the possibility to obtain active systems by means of well-assessed industrial processes, such extrusion. Del Nobile and colleagues compared the performances of synthetic recyclable systems (based on low density polyethylene) with biodegradable ones (PLA and PCL). They used an industrial process (twin-screw extruder) to prepare films and tested the effect of the manufacturing conditions on the antimicrobial effectiveness of films loaded with natural compounds. They found that the processing temperatures play a major role in determining the antimicrobial efficiency of the investigated active films. In particular, PLA and LDPE-based systems retained slight antimicrobial activity, whereas PCL ones, due to the much lower processing temperatures, allowed less degradation of antimicrobial activity.

Studies on the biodegradability of a blend of PCL with nonbiodegradable polymers were conducted by Iwamoto and Tokiwa. They found that PCL/polyolefin blends could also be a viable solution for food packaging because an enzymatic degradation of the blend, either PCL/LDPE or PCL/polypropylenede, can occur if viscosity and formulation are properly designed. Furthermore, they inferred from the phase structure of the blend that it is the continuous phase of PCL that is responsible for the degradability of the blend.

A more complex morphology for films is constituted by porous PCL films. They can be prepared by removing, typically by means of a specific solvent, a nonmiscible dispersed phase. The presence of porosity increases the surface exposed to enzymatic degradation and also increases permeability to gaseous or vapor species, both properly tailored by the proper choice of particle size distribution.

Reinforced PCL Composites

In general, materials based only on PCL are not used in applications where structural performances are required due to its limitations in glass transition temperature (< −60 °C) and elastic properties (Young's modulus around 0.5 GPa). Although it is a semicrystalline polymer, hence it can be used above its T_g, the elastic modulus is quite low when compared to other biodegradable polymers such PLA or polyhydroxybutirrate (PHB). A viable and effective method to improve the stiffness of PCL, and to increase its potential usability in load-bearing applications, is to reinforce it or, better, use one of its blends, with high aspect ratio fillers, such as short fibers or fabrics. In particular, since PCL is highly biodegradable, a lot of efforts have been put in investigating PCL composites reinforced with natural fibers.

The term natural fibers can point to either organic (flax, hemp, cellulose, ramie) or inorganic (hydroxyapatite, basalt, glass) fibers. They are very abundant in nature, can be biocompatible or bioresorbable, renewable and sustainable, and can allow reaching a high performance over cost ratio. In the following, the term natural fibers will be used to refer to organic fibers, which possess a much higher strength per unit weight than most inorganic fillers, lower density, and their biodegradable nature make natural fillers attractive as reinforcements for engineered biodegradable polymeric systems.

Since the high hydrophilicity of natural fibers, some drawbacks are present, such their incompatibility with the hydrophobic polymer matrix, the tendency to form aggregates during processing, and the poor resistance to moisture. These issues greatly reduce the potential of *as is* natural fibers as reinforcements, but specific surface treatments through physical and/or chemical processes for the improvement of fiber–matrix interaction can be exploited to improve fiber wetting and hence the structural performance of PCL composites.

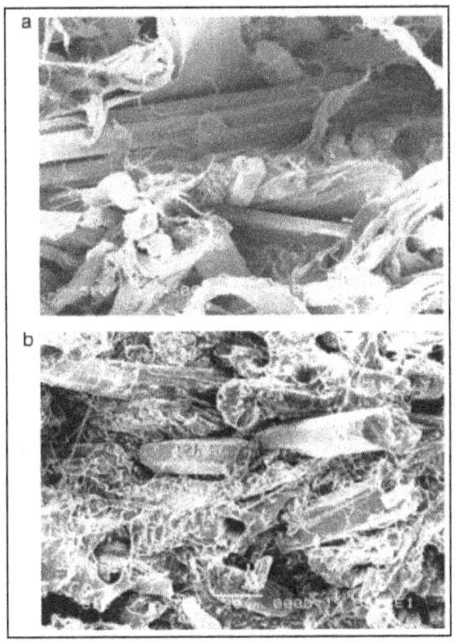

(a) Fracture surface of tensile specimens of flax/PCL composites. (b) Fracture surface of tensile specimens of flax/PCL-g-MA compatibilized composite.

Chemical modification of natural fibers was the first approach to increase the adhesion between the hydrophilic fibers and hydrophobic matrix. Even if the most promising way is to directly induce covalent bonds between fiber and matrix, the difficulties in finding affordable and controllable processes of this type leave room for other, simpler solutions. Alkali treatment is a well-known process; it is used from long time and allows to treat almost all the natural fibers with effective and good results. Another well-known treatment is with the use of maleated coupling agents, widely used for polyolefins because the interactions between the anhydride groups of maleated coupling agents and the hydroxyl groups of natural fibers can effectively improve the adhesion between fiber and matrix. Silane-coupling agents are also used, but they are not suitable for all polymers; and the need to use high temperatures to obtain a satisfying coupling. Coupling agents from natural resources have also been investigated such as lignin (successfully used in hemp and jute fiber composites), chitin and chitosan (used in wood-flour composites), fungus, or proteins (zein was used with very promising results).

Albeit the properties of natural fiber reinforced polymer composites are generally governed by the pretreatment process of fibers and the manufacturing process of the composites, the complexity and variability of fiber structures can lead to variability in the mechanical performances of composites with the use of the same fiber type in different production batches. Unlike for synthetic fibers, in fact, the fiber/matrix interface strength, the fiber impregnation, and the bonding can change along the natural fiber due to the uneven geometry or local composition. Furthermore, some processes are not targeted for natural fibers, but have been developed for the fast and reliable production of synthetic composites. Such issues should be taken into account in the design of a biodegradable composite because they can lead to lower than expected performances.

In structural applications often PCL is blended with other polymers with higher glass transition temperature. PLA/PCL composites were developed by Xu and co-workers by using ramie fiber by means of the *in situ* polymerization method. They treated the ramie fibers with a silane-coupling agent to increase the interfacial interactions with the polymeric blend. The compatibilized systems exhibited better mechanical properties (strength and impact resistance) with respect to composites prepared with untreated fibers. They also found that high ramie fibers content, and fibers as long as 6 mm had to be used to maximize performances.

Several works are present reporting on the mechanical behavior of composites based on PCL/starch blends reinforced with sisal fibers. In this case, the mechanical performance was improved after a specific alkaline surface treatment of sisal fibers. A commercial widely used biodegradable matrix based on PCL and starch was used (Mater-Bi-Z). The treatment improved the adhesion and the compatibility of the fibers and allowed a sound improvement of the elastic modulus and strength.

A PCL/starch blend has been used to prepare composites reinforced with kenaf and bamboo fibers, compatibilized with an alkali treatment by Shibata and others.

PCL/flax fiber composites were produced by Arbelaiz and colleagues by using maleated-coupling agents to prepare a poly(ε-caprolactone)-*grafted*-maleic anhydride copolymer (PCL-*g*-MA) as a compatibilizer intended to increase the interactions between matrix and fibers. The grafting reaction of maleic anhydride (MA) onto PCL polymer was carried out in the presence of dicumyl peroxide as an initiator. Composites fabricated with flax fiber bundles and PCL-*g*-MA matrix showed very high tensile and flexural strengths, as a consequence of the improved fiber/matrix adhesion. A potential drawback of using a compatibilizer is the effect on the crystallinity. In fact, both mechanical properties and crystallinity of composites decreased with the addition of small amounts of PCL-*g*-MA coupling agent. Crystallinity reduction was considered the possible reason for lower mechanical properties in some cases; hence, a proper increase of mechanical performance can be obtained by using proper amounts of compatibilizer, to maximize the interactions of the polymer with fibers and to recover some crystallinity.

The effect of fibers on the biodegradability can also arise. Natural fiber/polymer composites are usually intended for being used in nonload bearing, indoor, and low shelf life components due to their intrinsic vulnerability to environment. Nevertheless, some industrial fields require biodegradable composites to replace conventional, nondisposable ones. This implies that they have to be used in applications where such composites are exposed to moisture, thermal, fire, and ultraviolet degradation. Large fiber content provides strength to the polymer composite, but it also becomes an entry point for moisture attack. For this reason, several fiber treatments are developed to improve fiber/matrix interface and moisture durability. However, treated fibers were found to behave poorly when exposed to weather and UV stabilizers, and fire retardants must be used to increase outdoor usability and fire performance. These additives can compromise the strength of composites. A balance between strength and durability has to be accepted for using natural fibers in low demanding structural applications.

Natural composites based on PCL have been investigated by Wahit and co-workers, who reported that green composites with enhanced mechanical properties and biodegradation were prepared. Composites were prepared using various fiber treatments, and most of them proved to be effective in property enhancement, thereby raising the potential of natural fiber–based composites in outdoor applications. The optimum mechanical properties for most of the composites were obtained at an average fiber load of 30%. the use of fibers also allowed to reduce the cost of such green composites and hence expanding their application areas. Di Franco and co-workers produced composites with sisal fibers and PCL/starch matrix. The analysis of the hydrolytic stability of both composites and unfilled matrix showed that the use of fibers resulted in a promoted entrance of water, with a consequent swelling and hydrolysis of the starch phase. High fiber content should be reached to get a higher hydrolytic stability, but this probably was due to the presence of a fiber–fiber physical network. Microbial attack in biotic aqueous medium was evidenced by the presence of a biofilm, especially on the fiber surface. The presence of fibers indirectly promoted the degradation of PCL/starch in biotic environments, since they serve as support for the attack of the microorganisms and favored water entrance.

The thermal degradation behavior of composites made of TPS, PCL, and bleached sisal fibers was investigated by thermogravimetry analysis. The authors showed that at 5 wt% sisal fiber loading the apparent activation energy and the melting temperature of the composites decreased when compared with the TPS/PCL blend. A further increase of sisal fibers showed to increase the activation energy of composites thus improving their thermal stability.

Nylon

Nylon is the most useful synthetic material with applications varying from daily life activities to industries. It is a plastic which can be drawn into fibres or moulded into daily products for making amenities. We can live our entire life with nylon on our side. You hop across the nylon carpet to the kitchen, eat your breakfast on a nylon bowl after cleaning your teeth with a toothbrush whose bristles are made of nylon. A nylon umbrella over your head is used to move out of the house in heavy sunlight or to keep out of the rain.

The Science of Nylon

The term nylon points towards a polymer family known as linear polyamides. There are two approaches to making nylon for fibre applications. In the first approach, the molecules that consist of an acidic group (COOH) on every end react with molecules that contain amine (NH_2) groups at each end. The resulting nylon gets a name based on the number of carbon atoms that separate two amines and two acidic groups. Hence nylon 6,6 is widely used as fibres made from adipic acid and hexamethylene diamine.

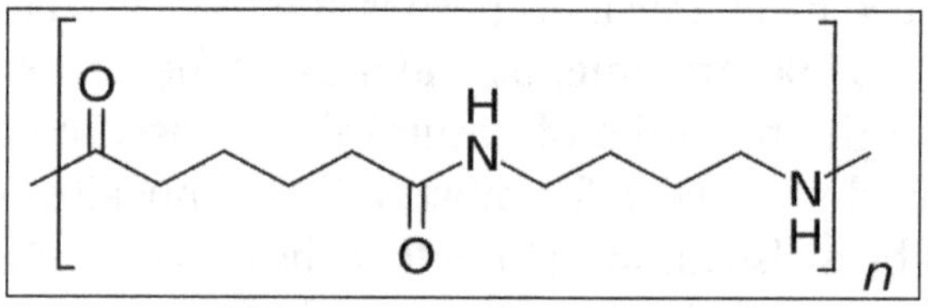

Structure of Nylon – 6, 6.

The salt which is formed by two compounds is known as nylon that has an exact ratio of 1:1 acid to base. This salt is dried and then heated under vacuum to remove water and form the polymer.

In the other approach, a compound that contains an amine at one end and acid at the other are polymerized to produce a chain with repeating units of (-NH-[CH_2]n-CO-)x. The nylon is referred to as nylon 6 if n = 5 which is another common form of this polymer. The commercial production of nylon 6 starts with caprolactam that use an open-ring polymerization.

In both the approaches, the polyamide is melt and drawn after cooling to obtain the desired properties of every intended use.

Properties of Nylon

- Lustrous.
- Elastic.
- Very strong.
- Damage resistant to oil and many chemicals.
- Resilient.
- Does not absorb water.
- Dries quickly.

Types of Nylon

- Nylon 6 – It was developed by Paul Schlack. It is formed by ring-opening polymerization.
- Nylon 510 – It is obtained from sebacic and pentamethylene diamine acid.

- Nylon 1,6 – It is produced from dinitriles with the help of acid catalysis.
- Nylon 66 – Wallace Carothers patented nylon 66 with the use of amide.

Uses of Nylon

- Clothing – Shirts, Foundation garments, lingerie, raincoats, underwear, swimwear and cycle wear.
- Industrial uses – Conveyer and seat belts, parachutes, airbags, nets and ropes, tarpaulins, thread, and tents.
- It is used to make a fish net.
- It is used as plastic in manufacturing machine parts.

Polyester

Polyester is a class of synthetic polymers built up from multiple chemical repeating units linked together by ester (CO-O) groups. Polyesters display a wide array of properties and practical applications. Permanent-press fabrics, disposable soft-drink bottles, compact discs, rubber tires, and enamel paints represent only a few of the products made from this group.

Polyesters most commonly are prepared from a condensation reaction between an organic alcohol (containing hydroxyl [OH] groups) and a carboxylic acid (containing carboxyl [COOH] groups). These two functional groups react to form the characteristic ester linkage, a chemical group with the structure:

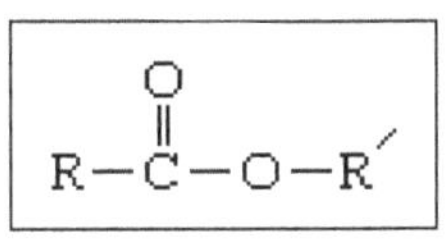

R and R′ represent the linked units that, repeated thousands of times within a single molecule, make up the long polymeric chain. The precise composition and structure of these repeating units vary widely, but roughly speaking they can be grouped into chains that are aliphatic (i.e., have an open structure) and those that contain ring-shaped molecular groups—particularly the large hydrocarbon aromatic groups.

Among the aliphatic group are the unsaturated polyesters, a class of resins that are molded into fibreglass-reinforced structures such as pleasure-boat hulls. Another aliphatic polyester is polyglycolic acid, a special type of degradable polymer that is made into bioabsorbable surgical sutures.

Ring-containing polyesters are the larger and commercially more important group. By far the most prominent member of this class is polyethylene terephthalate (PET), a stiff, strong polymer that is spun into fibres known by such trademarks as Dacron and Terylene. PET is also extruded into the film known as Mylar and is blow-molded into disposable beverage bottles. A related polyester is polybutylene terephthalate (PBT). PBT is used in applications similar to those of PET, and it is also employed in a synthetic rubber known as copolyester elastomer.

In general, the more aromatic groups included in the repeating units, the stiffer and higher-melting the polymer. This rule can be illustrated by polycarbonate, a rigid, tough, crystal-clear resin from which compact discs are made, and the polyarylates, a class of engineering plastics that often take the place of metals in machine parts.

Alkyd resins are oil-modified polyesters used in paints, varnishes, and other kinds of coating materials.

Uses and Applications

Fabrics woven or knitted from polyester thread or yarn are used extensively in apparel and home furnishings, from shirts and pants to jackets and hats, bed sheets, blankets, upholstered furniture and computer mouse mats. Industrial polyester fibers, yarns and ropes are used in car tire reinforcements, fabrics for conveyor belts, safety belts, coated fabrics and plastic reinforcements with high-energy absorption. Polyester fiber is used as cushioning and insulating material in pillows, comforters and upholstery padding. Polyester fabrics are highly stain-resistant—in fact, the only class of dyes which *can* be used to alter the color of polyester fabric are what are known as disperse dyes.

Polyesters are also used to make bottles, films, tarpaulin, canoes, liquid crystal displays, holograms, filters, dielectric film for capacitors, film insulation for wire and insulating tapes. Polyesters are widely used as a finish on high-quality wood products such as guitars, pianos and vehicle/yacht interiors. Thixotropic properties of spray-applicable polyesters make them ideal for use on open-grain timbers, as they can quickly fill wood grain, with a high-build film thickness per coat. Cured polyesters can be sanded and polished to a high-gloss, durable finish.

Industry

Basics

Polyester is a synthetic polymer made of purified terephthalic acid (PTA) or its dimethyl ester dimethyl terephthalate (DMT) and monoethylene glycol (MEG). With 18% market share of all plastic materials produced, it ranges third after polyethylene (33.5%) and polypropylene (19.5%).

The main raw materials are described as follows:

- Purified Terephthalic Acid (PTA) CAS-No.: 100-21-0:
  - Synonym: 1, 4 benzenedicarboxylic acid.
  - Sum formula: $C_6H_4(COOH)_2$, mol. weight: 166.13.
- Dimethylterephthalate (DMT) CAS-No.: 120-61-6:
  - Synonym: 1, 4 benzenedicarboxylic acid dimethyl ester.
  - Sum formula: C6H4(COOCH3)2, mol. weight: 194.19.
- Mono-ethylene Glycol (MEG) CAS No.: 107-21-1:
  - Synonym: 1, 2 ethanediol.
  - Sum formula: $C_2H_6O_2$, mol. weight: 62.07.

To make a polymer of high molecular weight a catalyst is needed. The most common catalyst is antimony trioxide (or antimony tri-acetate):

- Antimony Trioxide (ATO) CAS-No.: 1309-64-4:
  - Mol. weight: 291.51.
  - Sum formula: Sb_2O_3.

In 2008, about 10,000 tonnes Sb_2O_3 were used to produce around 49 million tonnes polyethylene terephthalate. Polyester is described as follows:

Polyethylene Terephthalate CAS-No.: 25038-59-9

Sum formula: H-[$C_{10}H_8O_4$]-n=60–120 OH, mol. unit weight: 192.17.

There are several reasons for the importance of polyester:

- The relatively easy accessible raw materials PTA or DMT and MEG.
- The very well understood and described simple chemical process of polyester synthesis.
- The low toxicity level of all raw materials and side products during polyester production and processing.
- The possibility to produce PET in a closed loop at low emissions to the environment.
- The outstanding mechanical and chemical properties of polyester.
- The recyclability.
- The wide variety of intermediate and final products made of polyester.

In the following table, the estimated world polyester production is shown. Main applications are textile polyester, bottle polyester resin, film polyester mainly for packaging and specialty polyesters for engineering plastics. According to this table, the world's total polyester production might exceed 50 million tons per annum before the year 2010.

Table: World polyester production by year.

Product type	2002 (million tonnes/year)	2008 (million tonnes/year)
Textile-PET	20	39
Resin, bottle/A-PET	9	16
Film-PET	1.2	1.5
Special polyester	1	2.5
Total	31.2	59

Raw Material Producer

The raw materials PTA, DMT, and MEG are mainly produced by large chemical companies which are sometimes integrated down to the crude oil refinery where *p*-Xylene is the base material to produce PTA and liquefied petroleum gas (LPG) is the base material to produce MEG.

Polyester Processing

After the first stage of polymer production in the melt phase, the product stream divides into two different application areas which are mainly textile applications and packaging applications. In the following table, the main applications of textile and packaging of polyester are listed.

Table: Textile and packaging polyester application list (melt or pellet).

Textile	Packaging
Staple fiber (PSF)	Bottles for CSD, water, beer, juice, detergents, etc.
Filaments POY, DTY, FDY	A-PET film
Technical yarn and tire cord	Thermoforming
Non-woven and spunbond	biaxial-oriented film (BO-PET)
Mono-filament	Strapping

A comparable small market segment (much less than 1 million tonnes/year) of polyester is used to produce engineering plastics and masterbatch.

In order to produce the polyester melt with a high efficiency, high-output processing steps like staple fiber (50–300 tonnes/day per spinning line) or POY /FDY (up to 600 tonnes/day split into about 10 spinning machines) are meanwhile more and more vertically integrated direct processes. This means the polymer melt is directly converted into the textile fibers or filaments without the common step of pelletizing. We are talking about full vertical integration when polyester is produced at one site starting from crude oil or distillation products in the chain oil → benzene → PX → PTA → PET melt → fiber/filament or bottle-grade resin. Such integrated processes are meanwhile established in more or less interrupted processes at one production site. Eastman Chemicals were the first to introduce the idea of closing the chain from PX to PET resin with their so-called INTEGREX process. The capacity of such vertically integrated production sites is >1000 tonnes/day and can easily reach 2500 tonnes/day.

Besides the above-mentioned large processing units to produce staple fiber or yarns, there are ten thousands of small and very small processing plants, so that one can estimate that polyester is processed and recycled in more than 10 000 plants around the globe. This is without counting all the companies involved in the supply industry, beginning with engineering and processing machines and ending with special additives, stabilizers and colors. This is a gigantic industry complex and it is still growing by 4–8% per year, depending on the world region.

Synthesis

Synthesis of polyesters is generally achieved by a polycondensation reaction. The general equation for the reaction of a diol with a diacid is:

$$(n + 1)\ R(OH)_2 + n\ R'(COOH)_2 \rightarrow HO[ROOCR'COO]_nROH + 2n\ H_2O$$

Azeotrope Esterification

In this classical method, an alcohol and a carboxylic acid react to form a carboxylic ester. To assemble a polymer, the water formed by the reaction must be continually removed by azeotrope distillation.

Alcoholic Transesterification

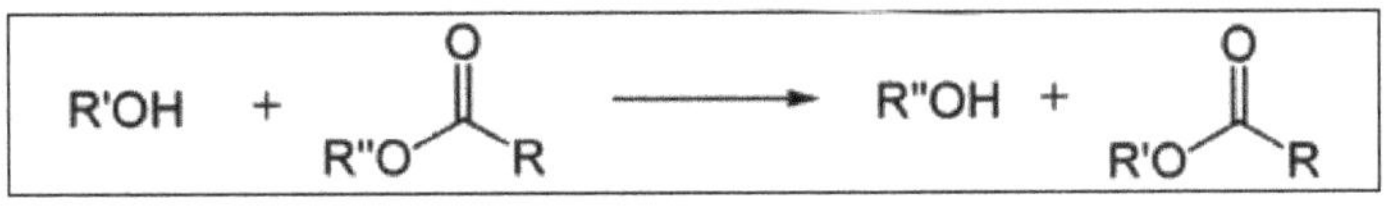

Transesterification: An alcohol-terminated oligomer and an ester-terminated oligomer condense to form an ester linkage, with loss of an alcohol. R and R' are the two oligomer chains, R" is a sacrificial unit such as a methyl group (methanol is the byproduct of the esterification reaction).

Acylation (HCl Method)

The acid begins as an acid chloride, and thus the polycondensation proceeds with emission of hydrochloric acid (HCl) instead of water. This method can be carried out in solution or as an enamel.

Silyl Method

In this variant of the HCl method, the carboxylic acid chloride is converted with the trimethyl silyl ether of the alcohol component and production of trimethyl silyl chloride is obtained.

Acetate Method (Esterification)

Ring-opening Polymerization

Aliphatic polyesters can be assembled from lactones under very mild conditions, catalyzed anionically, cationically or metallorganically. A number of catalytic methods for the copolymerization of epoxides with cyclic anhydrides have also recently been shown to provide a wide array of functionalized polyesters, both saturated and unsaturated.

Polystyrene

Polystyrene is a hard, stiff, brilliantly transparent synthetic resin produced by the polymerization of styrene. It is widely employed in the food-service industry as rigid trays and containers, disposable eating utensils, and foamed cups, plates, and bowls. Polystyrene is also copolymerized, or blended with other polymers, lending hardness and rigidity to a number of important plastic and rubber products.

Styrene is obtained by reacting ethylene with benzene in the presence of aluminum chloride to yield ethylbenzene. The benzene group in this compound is then dehydrogenated to yield phenylethylene, or styrene, a clear liquid hydrocarbon with the chemical structure CH_2=CHC_{6H_5}. Styrene is polymerized by using free-radical initiators primarily in bulk and suspension processes, although solution and emulsion methods are also employed. The structure of the polymer repeating unit can be represented as:

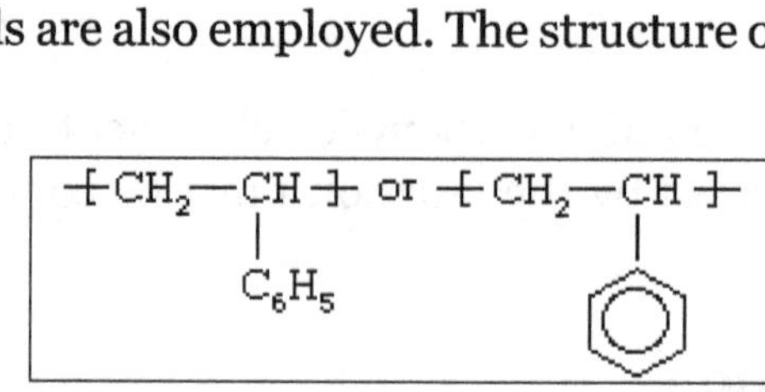

The presence of the pendant phenyl (C_6H_5) groups is key to the properties of polystyrene. Solid polystyrene is transparent, owing to these large, ring-shaped molecular groups, which prevent the polymer chains from packing into close, crystalline arrangements. In addition, the phenyl rings restrict rotation of the chains around the carbon-carbon bonds, lending the polymer its noted rigidity.

The polymerization of styrene has been known since 1839, when German pharmacist Eduard Simon reported its conversion into a solid later named metastyrol. As late as 1930 little commercial use had been found for the polymer because of brittleness and crazing (minute cracking), which were caused by impurities that brought about the cross-linking of the polymer chains. By 1937 American chemist Robert Dreisbach and others at the Dow Chemical Company's physics laboratory had obtained purified styrene monomer through the dehydrogenation of ethylbenzene and developed a pilot polymerization process. By 1938 polystyrene was being produced commercially. It quickly became one of the most important modern plastics, owing to the low cost of producing large volumes of styrene monomer, the ease of shaping the melted polymer in injection-molding operations, and the optical and physical properties of the material.

Polystyrene foam was formerly made with the aid of chlorofluorocarbon blowing agents—a class of compounds that has been banned for environmental reasons. Now foamed by pentane or carbon dioxide gas, polystyrene is made into insulation and packaging materials as well as food containers such as beverage cups, egg cartons, and disposable plates and trays. Solid polystyrene products include injection-molded eating utensils, videocassettes and audiocassettes, and cases for audiocassettes and compact discs. Many fresh foods are packaged in clear vacuum-formed polystyrene trays, owing to the high gas permeability and good water-vapour transmission of the material. The clear windows in many postage envelopes are made of polystyrene film. The plastic recycling code number of polystyrene is #6. Recycled polystyrene products are commonly melted down and reused in foamed insulation.

Despite its advantageous properties, polystyrene is brittle and flammable; it also softens in boiling water and, without the addition of chemical stabilizers, yellows upon prolonged exposure to sunlight. In order to reduce brittleness and improve impact strength, more than half of all polystyrene produced is blended with 5 to 10 percent butadiene rubber. This blend, suitable for toys and appliance parts, is marketed as high-impact polystyrene (HIPS).

Semi-synthetic Polymers

Semi synthetic polymers are derived from naturally occurring polymers by chemical modifications.

Examples: Vulcanized rubber, gun cotton and cellulose diacetate.

Vulcanized rubber is used in making tyres. Gun cotton which is cellulose nitrate is used in making explosives. Cellulose on acetylation with acetic anhydride in the presence of sulphuric acid forms cellulose diacetate used in making threads and materials like films, glasses etc.

Classification based on Structure of Polymers

Linear Polymers

A polymer that has only a minimal amount of branching or no branching at all. A linear polymer is a macromolecule made out of many monomer units arranged in a straight line. A linear polymer consists of a single continuous chain of repeat units. This chain is called the backbone of the polymer. This liner polymer can have side groups attached to the backbone. These side groups are called pendant groups. But these side groups are not side chains.

In a linear polymer, the pendant groups can be arranged in different patterns. These patterns are described under the concept of tacticity – the relative regularity of a polymer chain. According to the tacticity of a polymer, these linear polymers can be divided into three main groups as isotactic polymers, syndiotactic polymers, and atactic polymers. Isotactic polymers have their pendant groups on the same side of the polymer chain. Syndiotactic polymers have their pendant groups in an alternating pattern. Atactic polymers have the pendant groups in a random manner.

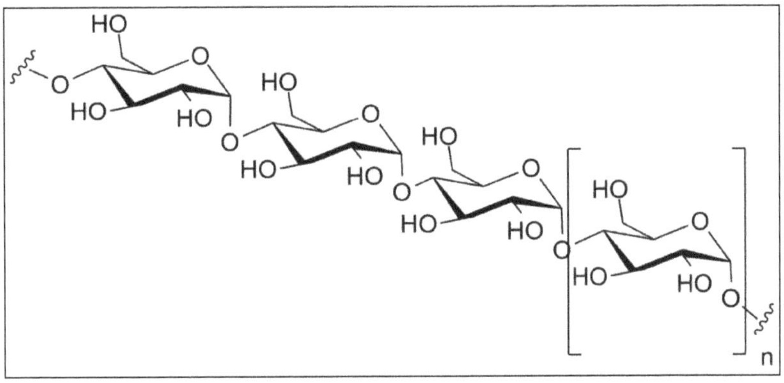

Amylose is a Linear Polymer of Glucose Monomers.

The backbone can be made out of either the same monomer or from different monomers. If it is the same polymer, it is called a linear homopolymer. If the backbone is made out of different monomers, it is called a linear copolymer. These copolymers can be found in different forms such as alternating copolymers (where the polymer chain is composed of regular alternating monomers), periodic copolymers (where the monomers are arranged in a repeating sequence) and, block copolymers (where the blocks of different monomers are arranged in a linear chain).

These macromolecules are closely packed due to less steric hindrance. This gives the linear polymers a high density, high tensile strength, and a high melting point.

Branch Chain Polymers

A branched polymer is one of the structural classifications of a polymer. Others include Linear Polymers, Cross-linked polymers, and Crystalline or Amorphous polymers.

Typically, a branched polymer is formed when the growth point is moved during the polymerization process. This means that a substituent (e.g. a hydrogen atom) on a monomer subunit of the main polymer is replaced covalently by another bonded chain of the polymer. Branching may, therefore, result from the formation of carbon-carbon or various other types of covalent bonds. Branching may also occur by ester and amide bonds. this is typical via a condensation reaction, producing one molecule of water (or HCl) for each bond formed.

Since the properties of a polymer are strongly affected by their molecular weight and their chain structure. Branching effects on a polymer depend on the number and length of the branches. For example, short branches they reduce the amount of crystallinity whereas long branches undergo side chain crystallization and are able to form lamellar crystals.

Branched polymers have more compact and symmetrical molecular conformations and even a small amount of long-chain branches can significantly improve the mechanical properties of the polymer. The degree of branching and the length of the branches also affect the density of the polymer. Typically, the higher the density of the polymer the higher the degree of crystallinity and the stiffer, harder, and stronger the polymer.

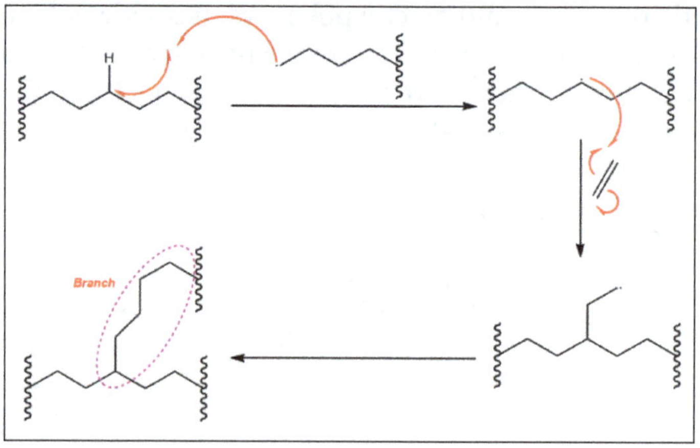

Cross Linked or Network Polymers

Crosslinked polymers are macromolecules that has covalent bonds between polymer molecules. A crosslink is a bond between two polymer chains. These crosslinks can be either ionic bonds or covalent bonds. These crosslinks can be formed either during the process of polymerization or after polymerization.

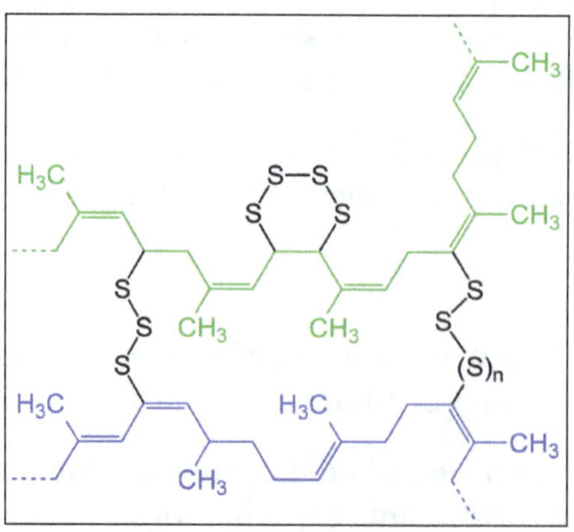

Vulcanization of Rubber.

Since crosslinks between polymer chains are stronger than normal intermolecular attractions, crosslinking forms a stable and stronger polymer material. Crosslinked polymers can be found in both synthetic polymers and as naturally occurring polymers. Crosslinks can be formed from chemical reactions using crosslinking reagents. The most common example of crosslinked polymers is the vulcanized rubber. Since natural rubber is not enough stiff and rigid, rubber is vulcanized. It is heated with sulfur, so sulfur molecules form covalent bonds in the rubber polymer chains, connecting the chains with each other. This makes rubber a stiff, rigid material and durable.

The degree of crosslinking gives the number of crosslinking per mole of a material. The degree of crosslinking is usually measured by swelling experiment. Here, the material is placed in a container with a suitable solvent. Then the change of mass or the change of volume is measured. If the degree of crosslinking is low, the material swells more.

Classification based on Mode of Polymerisation

Addition Polymers

An addition polymer is a polymer formed by chain addition reactions between monomers that contain a double bond Condensation polymers are examples of A-A polymers.

Molecules of ethene can polymerize with each other under the right conditions to form the polymer called polyethylene.

$$nCH_2 = CH_2 \rightarrow (CH_2CH_2)_n -$$

The letter n stands for the number of monomers that are joined in repeated fashion to make the polymer and can have a value in the hundreds or even thousands.

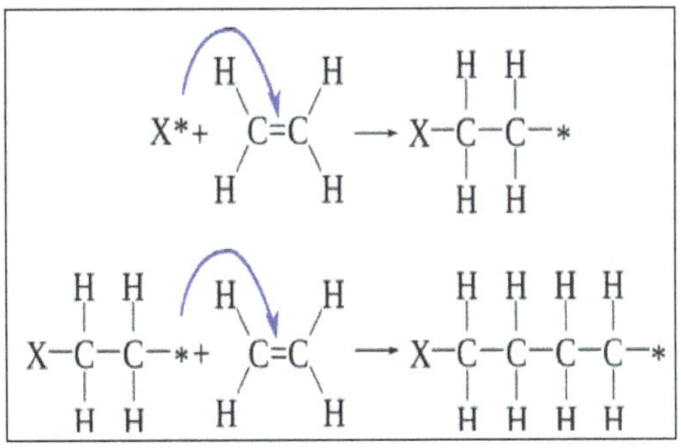

Polyethylene synthesis.

The reactions above show the basic steps to form an addition polymer.

- Initiation - a free radical initiator (X∗) attacks the carbon-carbon double bond (first step above). The initiator can be something like hydrogen peroxide. This material can easily split to form two species with a free electron attached to each: H–O–O–H→2H–O·. This free radical attacks a carbon-carbon double bond. One of the pi electrons forms a single bond with the initiator while the other pi electron forms a new free radical on the carbon atom.

- Propagation - the new free radical compound interacts with another alkane, continuing the process of chain growth.
- Termination occurs whenever two free radicals come in contact with one another. The two free electrons form a covalent bond and the free radical on each molecule no longer exists.

Polyethylene can have different properties depending on the length of the polymer chains and on how efficiently they pack together. Some common products made from different forms of polyethylene include plastic bottles, plastic bags, and harder plastic objects such as milk crates.

Several other kinds of unsaturated monomers can be polymerized and are components in common household products. Polypropylene is stiffer than polyethylene and is in plastic utensils and some other types of containers.

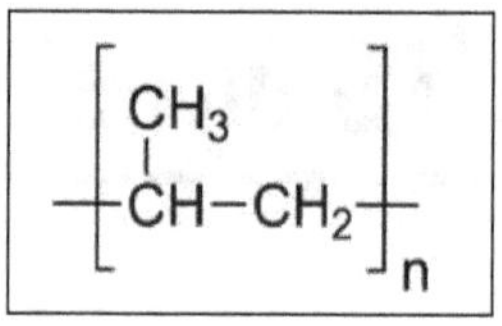

Polypropylene structure.

Polystyrene is used in insulation and in molded items such as coffee cups.

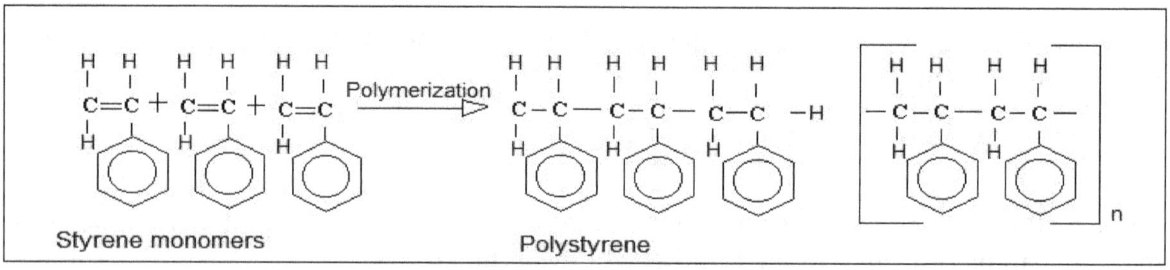

Polystyrene synthesis and structure.

Polyvinyl chloride (PVC) is extensively used for plumbing pipes.

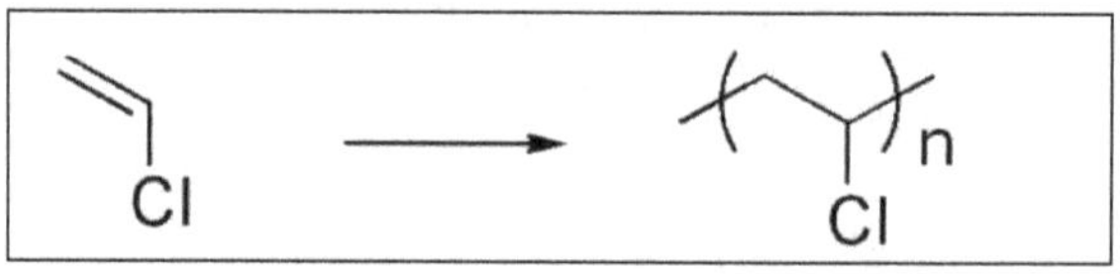

Polyvinyl chloride.

Polyisoprene is a polymer of isoprene and is better known as rubber. It is produced naturally by rubber trees, but several variants have been developed which demonstrate improvements on the properties of natural rubber.

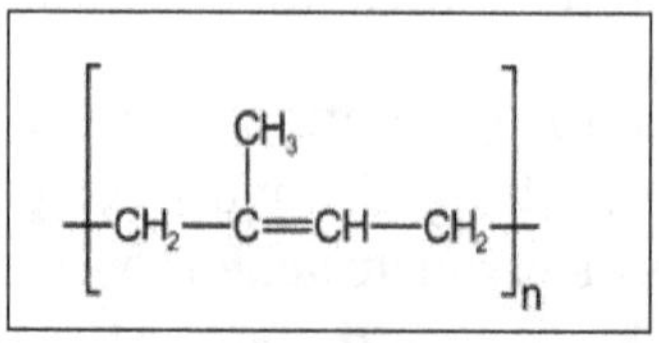

Polyisoprene.

Condensation Polymers

Condensation polymers are formed when lots of smaller molecules (monomers) join together to form a larger molecule (polymer), resulting in the loss of small molecules. Condensation polymers are examples of A-B polymers. The polyester Dacron and the polyamide Nylon 66, shown here, are two examples of synthetic condensation polymers, also known as step-growth polymers. In contrast to chain-growth polymers, most of which grow by carbon-carbon bond formation, step-growth polymers generally grow by carbon-heteroatom bond formation (C-O & C-N in Dacron & Nylon respectively). Although polymers of this kind might be considered to be alternating copolymers, the repeating monomeric unit is usually defined as a combined moiety.

Examples of naturally occurring condensation polymers are cellulose, the polypeptide chains of proteins, and poly(β-hydroxybutyric acid), a polyester synthesized in large quantity by certain soil and water bacteria.

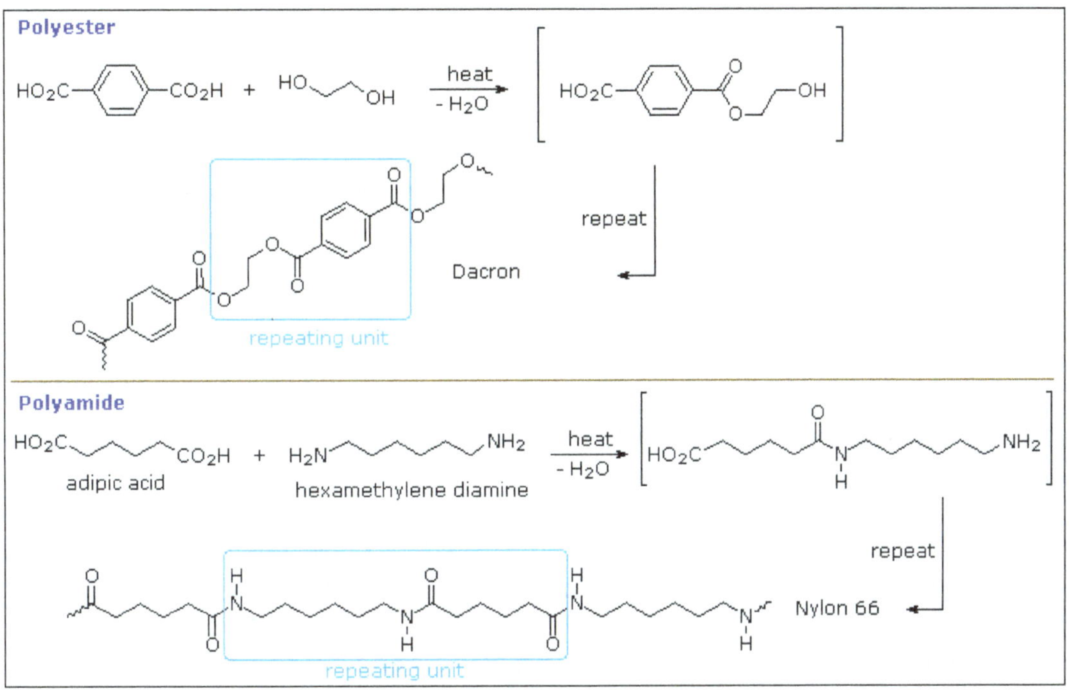

Examples of condensation polymers.

Characteristics of Condensation Polymers

Condensation polymers form more slowly than addition polymers, often requiring heat, and they are generally lower in molecular weight. The terminal functional groups on a chain remain active, so that groups of shorter chains combine into longer chains in the late stages of polymerization. The presence of polar functional groups on the chains often enhances chain-chain attractions, particularly if these involve hydrogen bonding, and thereby crystallinity and tensile strength. The following examples of condensation polymers are illustrative.

Note that for commercial synthesis the carboxylic acid components may actually be employed in the form of derivatives such as simple esters. Also, the polymerization reactions for Nylon 6 and

Spandex do not proceed by elimination of water or other small molecules. Nevertheless, the polymer clearly forms by a step-growth process.

Table: Some Condensation Polymers.

Formula	Type	Components	T_g ºC	T_m ºC
$\sim[CO(CH_2)_4CO\text{-}OCH_2CH_2O]_n\sim$	**polyester**	$HO_2C\text{-}(CH_2)_4\text{-}CO_2H$ $HO\text{-}CH_2CH_2\text{-}OH$	<0	50
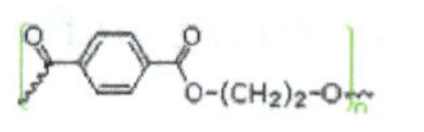	**polyester** Dacron, Mylar	para $HO_2C\text{-}C_6H_4\text{-}CO_2H$ $HO\text{-}CH_2CH_2\text{-}OH$	70	265
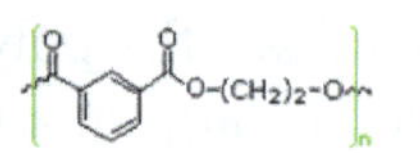	**polyester**	meta $HO_2C\text{-}C_6H_4\text{-}CO_2H$ $HO\text{-}CH_2CH_2\text{-}OH$	50	240
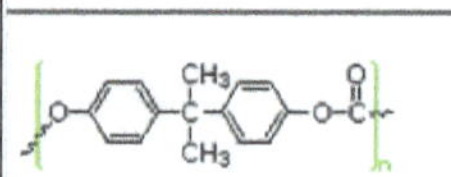	**polycarbonate** Lexan	$(HO\text{-}C_6H_4\text{-})_2C(CH_3)_2$ (Bisphenol A) $X_2C{=}O$ ($X = OCH_3$ or Cl)	150	267
$\sim[CO(CH_2)_4CO\text{-}NH(CH_2)_6NH]_n\sim$	**polyamide** Nylon 66	$HO_2C\text{-}(CH_2)_4\text{-}CO_2H$ $H_2N\text{-}(CH_2)_6\text{-}NH_2$	45	265
$\sim[CO(CH_2)_5NH]_n\sim$	**polyamide** Nylon 6 Perlon	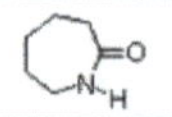	53	223
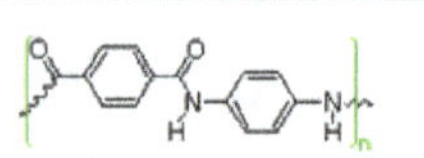	**polyamide** Kevlar	para $HO_2C\text{-}C_6H_4\text{-}CO_2H$ para $H_2N\text{-}C_6H_4\text{-}NH_2$	---	500
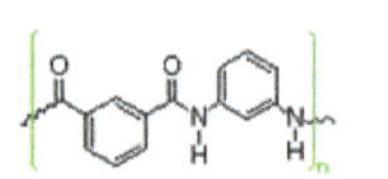	**polyamide** Nomex	meta $HO_2C\text{-}C_6H_4\text{-}CO_2H$ meta $H_2N\text{-}C_6H_4\text{-}NH_2$	273	390

The difference in Tg and Tm between the first polyester (completely aliphatic) and the two nylon polyamides (5th & 6th entries) shows the effect of intra-chain hydrogen bonding on crystallinity. The replacement of flexible alkylidene links with rigid benzene rings also stiffens the polymer chain, leading to increased crystalline character, as demonstrated for polyesters (entries 1, 2 &3) and polyamides (entries 5, 6, 7 & 8). The high Tg and Tm values for the amorphous polymer Lexan are consistent with its brilliant transparency and glass-like rigidity. Kevlar and Nomex are extremely tough and resistant materials, which find use in bullet-proof vests and fire resistant clothing.

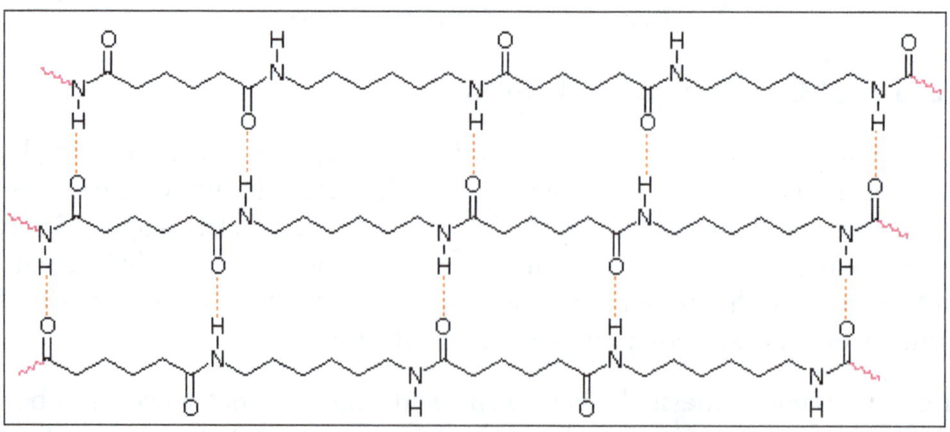

Interchain hydrogen bonding enhances crystallinity.

Many polymers, both addition and condensation, are used as fibers The chief methods of spinning synthetic polymers into fibers are from melts or viscous solutions. Polyesters, polyamides and polyolefins are usually spun from melts, provided the Tm is not too high. Polyacrylates suffer thermal degradation and are therefore spun from solution in a volatile solvent. Cold-drawing is an important physical treatment that improves the strength and appearance of these polymer fibers. At temperatures above T_g, a thicker than desired fiber can be forcibly stretched to many times its length; and in so doing the polymer chains become untangled, and tend to align in a parallel fashion. This cold-drawing procedure organizes randomly oriented crystalline domains, and also aligns amorphous domains so they become more crystalline. In these cases, the physically oriented morphology is stabilized and retained in the final product. This contrasts with elastomeric polymers, for which the stretched or aligned morphology is unstable relative to the amorphous random coil morphology.

This cold-drawing treatment may also be used to treat polymer films (e.g. Mylar & Saran) as well as fibers.

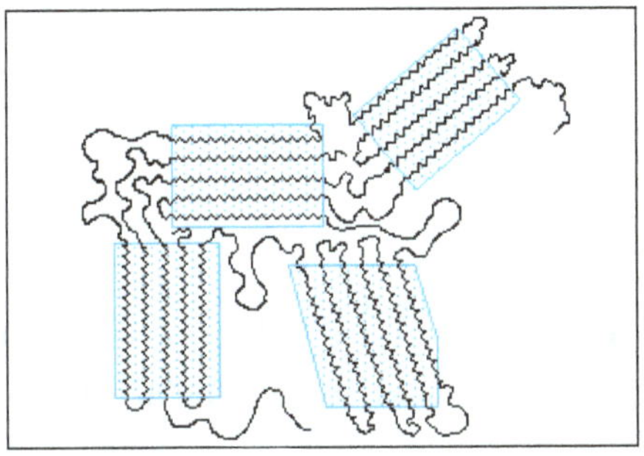

Random crystalline domains.

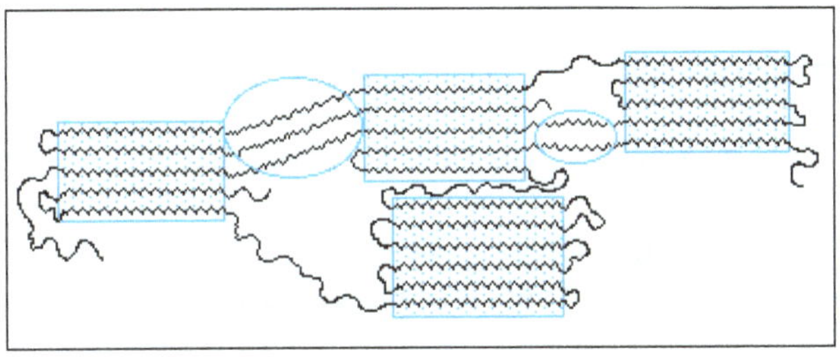

Ordered crystalline domains.

Step-growth polymerization is also used for preparing a class of adhesives and amorphous solids called epoxy resins. Here the covalent bonding occurs by an S_N2 reaction between a nucleophile, usually an amine, and a terminal epoxide. In the following example, the same bisphenol A intermediate used as a monomer for Lexan serves as a difunctional scaffold to which the epoxide rings are attached. Bisphenol A is prepared by the acid-catalyzed condensation of acetone with phenol.

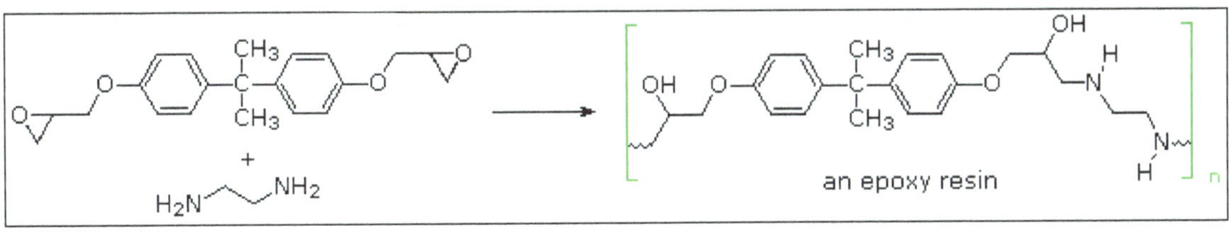

Classification based on Molecular Forces

Elastomers

Elastomers are polymers that are having viscosity as well as elasticity and therefore are known as viscoelasticity. The molecules of elastomers held together by weak inter-molecular forces, generally they exhibit low Young's modulus and high yield strength or high failure strain. They inherit the unique property of regaining original shape and size after being stretched to a great extent.

Examples of Elastomers

Following are the examples of elastomers with their applications:

- Natural rubber: These are used in the automotive industry and in the manufacture of medical tubes, balloons, adhesives.
- Polyurethanes: These are used in the textile industry for manufacturing elastic clothing like lycra.
- Polybutadiene: These are used for providing wear resistance in wheels of vehicles.
- Silicone: These are used in the manufacture of medical prostheses, lubricants as they have an excellent chemical and thermal resistance.
- Neoprene: These are used in the manufacture of wet-suits and in industrial belts.

Properties of Elastomers

Following are the properties of elastomers:

- Temperature: The specific working temperature of elastomers vary depending on the factors like media compatibility, seal design, and the dynamic and static operation.
- Low-temperature flexibility: The rate of recovery of elastomeric material can be studied by subjecting the material to low-temperature retraction.
- Hardness: The measurement of the material's resistance towards the deforming force for a defined length of time is done by measuring the hardness. It differs from material to material. The soft compounds deform easily and have high friction while the harder compounds have high resistance and low friction.
- Ageing: This property helps to understand the behavior of a material when exposed to heat. If the elastomers are pushed beyond their aging resistance, they will suffer from hardening, cracking and splitting.
- Colour: Colouring is used mainly to differentiate between the compound grades based on their usage.
- Elongation at break: This property is used for testing the moment of rupture when the material is under tensile stress.

Types of Elastomers

Following are the two types of elastomers:

- Saturated Elastomers: Sulphur vulcanization can not cure them. They showcase superior stability against oxygen, radiation, heat, and ozone. Comparably they are less reactive.

Their reactivity is limited to certain circumstances and condition. Polyacrylic rubber and silicone rubber are examples of saturated elastomers.

- Unsaturated Elastomers: They can be cured with Sulphur vulcanization process. Butyl rubber and natural polyisoprene are examples of unsaturated elastomers.

Examples of Elastomers

They play an essential and ubiquitous role in day to day life due to their elasticity, flexibility, insolubility and lot many prominent features. Some of their applications are listed below:

- Motor vehicles: Some elastomers like thermoset doesn't melt easily making it efficient in building seals, tyres various components throughout the automobile design. Especially those components which will be exposed to heat during the functioning. The material of the type polybutadiene offers extraordinary wear resistance hence they are preferable in building tyres.
- Consumer products: This comprises the widest range of products starting from shoe sole to baby pacifiers and many more miscellaneous.
- Constructions: Adhesives and sealants materials enfolded under elastomers, which are an unavoidable part of any constructions. Especially for filling the gaps.
- Industrial products: Elastomers are hugely used in making industrial tools, appliances, belts, molds, lubricants, etc.
- Wire and cable: Material needed to build wires should have high resistance to heat, easily reshaped(elongated) and provide insulation. The elastomers like neoprene are perfect for this.
- Medical products: Medical field needs a wide range of products like prosthetics, lubricants, molds with superior class of chemical and thermal resistance. Elastomer like silicon has widely used the material to build them and many other goods.

Thermosetting

A thermosetting polymer which is also known as a thermoset or thermosetting plastic is a polymer consisting of cross-linked structure or heavily branched molecules.

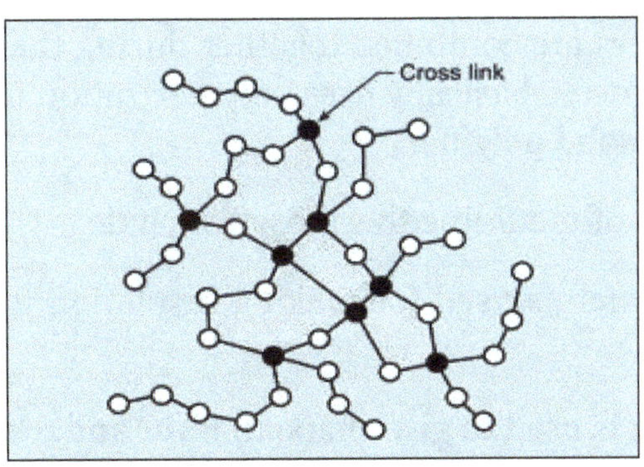

Thermosetting Polymers Structure.

These polymers which are in the soft solid or viscous state on heating undergo extensive cross-linking in moulds and become irreversibly hard as well as insoluble products.

Properties of Thermosetting Polymers

- One of the main properties of thermoset plastics or polymers is that they harden during the moulding process and after solidifying they cannot be softened.
- Typically, when the polymers are moulded and shaped they acquire a three-dimensional cross-linked structure along with strong covalent bonds that further help them retain their strength as well as a structure even if the temperatures are set high.
- However, thermoset plastics are brittle and tend to char and burn if heat is applied for a prolonged state. Normally, thermosets decompose before melting. Meanwhile, thermoset resins are insoluble in most of the inorganic solvents.

Thermosetting Process

Processing thermoset polymers usually involve three stages:

- The first stage is commonly referred to as resole where during this stage the resin is an insoluble and fusible condition or state.
- In the second stage, the thermosets are only partly soluble and they tend to display the characteristics of a thermoplastic where the changes are reversible.
- However, this is a temporary state as the molten form of thermoset lasts only for a short time because the increase in temperatures promotes cross-linking.
- The final stage is where the cross-linking reaction occurs and the final structure of thermosets is constructed. This mostly involves the moulding phase which is conducted under controlled temperature and some application of pressure.
- The final product will have a three-dimensional internal network structure consisting of highly cross-linked polymer chains. At this point, the product cannot be thermally deformed.

In addition, thermoset plastics can also be processed via a single-stage method called reaction injection moulding (RIM). While this process is similar to the method described above the only distinction here is that polymers are combined together during the curing process to form a permanent chemical bond. The process basically uses polymerization in the mould instead of cooling to facilitate the formation of a solid polymer.

Some of the common methods of moulding thermosets include:

- Extrusion moulding which is used for making insulation for electrical cables, threads of fabric and pipes.
- Compression moulding is used to give shape to BMC and SMC thermosetting plastics.

- Spin casting which is a process mainly used for the production of figurines, fishing lures and jigs, emblems and even replacement parts.

Examples:

- One of the most common examples of thermosets is bakelite which is relatively a bad conductor of electricity and heat.
- It is mainly used for making electrical switches, handles of various utensils, etc.
- Another example is Melamine which has a capacity to resist fire and heat much efficiently than other plastics.
- It is used in kitchenware and fabrics as well as floor tiles.
- Some other examples of thermoset plastic polymers include silicones, vulcanized rubber, epoxies, polyesters and phenolics.

References

- Rubber: chemistrylearning.com, Retrieved 29 June, 2019
- Biological-polymers: thoughtco.com, Retrieved 14 July, 2019
- Rosato, Dominick V.; Rosato, Donald V.; Rosato, Matthew V. (2004). Plastic product material and process selection handbook. Elsevier. P. 85. ISBN 978-1-85617-431-2
- Synthetic-polymers, chemistry: byjus.com, Retrieved 11 January, 2019
- Polyethylene, science: britannica.com, Retrieved 18 April, 2019
- Mendelson, Cheryl (17 May 2005). Home Comforts: The Art and Science of Keeping House. Simon and Schuster. ISBN 9780743272865
- Polyvinyl-chloride-pvc-plastic, selection-guide: omnexus.specialchem.com, Retrieved 27 May, 2019
- Polyester, science: britannica.com, Retrieved 21 July, 2019
- Branched-polymer, glossary: chemistryscore.com, Retrieved 13 July, 2019

3

Inorganic and Organic Polymers

Organic polymers contain carbon atoms and inorganic polymers have ionic and hydrocarbon bonds but do not contain carbon atoms. Olysiloxanes, polyphosphazenes, polysilanes, etc. are some inorganic polymers and polytetrafluoroethylene, fluoropolymers, etc. are a few organic polymers. This chapter discusses about these organic and inorganic polymers in detail.

Inorganic Polymers

Inorganic polymer isany of a class of large molecules that lack carbon and are polymers—that is, made up of many small repeating units called monomers. Nature abounds with carbon-based (that is, organic) polymers, such as wool, silk, proteins, starch, and cellulose. In addition, rubber and plastic are made of a wide variety of man-made organic polymers. But many inorganic compounds, such as oxyacids and oxy-anions, also form polymers. This is especially true of weak acids, such as boric acid, H_3BO_3, and silicic acid, H_4SiO_4. In the anions of weak acids, a high density of negative charge resides on the oxygen atoms. This charge density can be reduced by the process of polymerization.

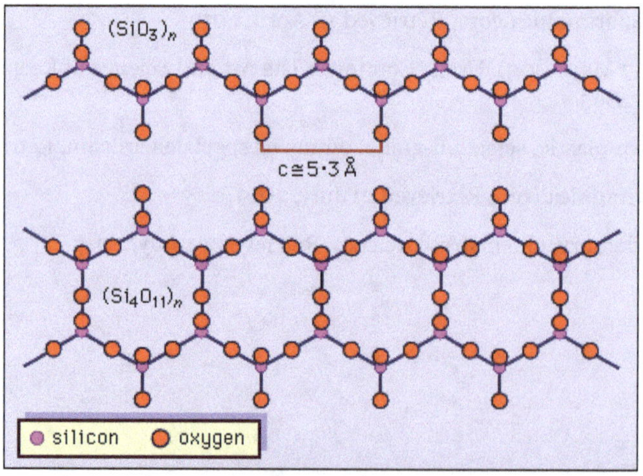

The single-chain silicon-oxygen tetrahedral structure (SiO_3)n of pyroxene minerals and the double-chain structure (Si4011)n of amphibole minerals are examples of inorganic polymers of silicon.

Borates

These compounds are salts of the oxyacids of boron (B), such as boric acid, H_3BO_3, metaboric acid, HBO_2, and tetraboric acid, $H_2B_4O_7$. Borates result either from the reaction of a base with a boron oxyacid or from the melting of boric acid or boron oxide, B_2O_3, with a molten metal oxide or hydroxide. Borate anion structures range from the simple trigonal planar BO_3^{3-} ion to rather complex structures containing chains and rings of three- and four-coordinated boron atoms. For example, calcium metaborate, CaB_2O_4, consists of infinite chains of $B_2O_4^{2-}$ units, whereas potassium borate, $K[B_5O_6(OH)_4] \cdot 2H_2O$ (commonly written as $KB_5O_8 \cdot 4H_2O$), consists of two B_3O_3 rings linked through a common four-coordinated boron atom. The tetraborates, $B_4O_5(OH)_4^{2-}$, contain both three- and four-coordinated boron surrounded trigonally and tetrahedrally, respectively, by oxygen (O) atoms. Commercially, the most important borate is borax, or sodium tetraborate decahydrate, $Na_2B_4O_7 \cdot 10H_2O$. Borax is found naturally in dry lake beds, such as Searles Lake in California. It can be used to soften water and to make washing compounds. Its usefulness arises from the insolubility of calcium and magnesium borates and the alkaline or basic nature of aqueous solutions of borax. Borax is also used in the manufacture of borosilicate glass and enamels and as a fire retardant.

Silicates

Silicates are salts containing anions of silicon (Si) and oxygen. There are many types of silicates, because the silicon-to-oxygen ratio can vary widely. In all silicates, however, silicon atoms are found at the centres of tetrahedrons with oxygen atoms at the corners. The silicon is always tetravalent (i.e., has an oxidation state of +4). The variation in the silicon-to-oxygen ratio occurs because the silicon-oxygen tetrahedrons may exist as discrete, independent units or may share oxygen atoms at corners, edges, or—in rarer instances—faces in several ways. Thus, the silicon-to-oxygen ratio varies according to the extent to which the oxygen atoms are shared by silicon atoms as the tetrahedrons are linked together. The linkage of these tetrahedrons provides a rather convenient way of classifying silicates. Seven different classifications are commonly recognized.

- In some silicates, individual SiO_4^{4-} tetrahedrons exist as independent units. Silicates of magnesium (Mg_2SiO_4) and zirconium ($ZrSiO_4$) are examples.
- Two SiO_4 tetrahedrons share one corner oxygen atom to form discrete $Si_2O_7^{6-}$ ions. Two compounds with this type of linkage are $Ca_2ZnSi_2O_7$ and $Zn_4(OH)_2Si_2O_7 \cdot H_2O$.
- SiO_4 tetrahedrons may share corners and form closed rings. In $BaTiSi_3O_9$, three SiO_4 tetrahedrons share corners, whereas in $Be_3Al_2Si_6O_{18}$ (beryl, the deep green variety of which is known as emerald), six tetrahedrons share corners to form a closed ring.
- SiO_4 tetrahedrons in which each tetrahedron shares two oxygen atoms from two other tetrahedrons exist as chains in some silicates. An example of this type of silicate is $CaMg(SiO_3)_2$. From the formula it appears that SiO_3^{2-} ions exist, but these ions do not occur as independent entities. Parallel chains extend the full length of the crystal and are held together by the positively charged metal ions lying between them.
- When SiO_4 tetrahedrons in single chains share oxygen atoms, double silicon-oxygen chains form. Metal cations link the parallel chains together. Many of these silicates are fibrous in nature, because the ionic bonds between the metal cations and the silicate anions are not as strong as the silicon-oxygen bonds within the chains. A class of fibrous silicate minerals that belong to this group is collectively called asbestos. The best known and most abundant kind of asbestos is chrysotile, which has the formula $Mg_3(Si_2O_5)(OH)_4$. This compound

exists as fibres more than 20 mm (0.8 inch) in length. It was used in the past in many fireproofing and insulation applications, but its use for these purposes has been discontinued because prolonged exposure to airborne asbestos fibres may cause lung cancer.

- When oxygen atoms are shared between double chains, silicon-oxygen sheets are formed. Metal ions form ionic bonds between the sheets. These ionic bonds are weaker than the silicon-oxygen bonds within the sheets, so silicates with this structure cleave into thin layers. An example of this class of silicates includes talc, $Mg_3Si_4O_{10}(OH)_2$.
- A most interesting class of silicates consists of the zeolites. These compounds are three-dimensional silicon-oxygen networks with some of the tetravalent silicon ions replaced by trivalent aluminum (Al^{3+}) ions. The negative charge that results—because each Al^{3+} ion has one fewer positive charge than the Si^{4+} ion it replaces—is neutralized by a distribution of positive ions throughout the network. An example of a zeolite is $Na_2(Al_2Si_3O_{10}) \cdot 2H_2O$. Zeolites are characterized by the presence of tunnels and systems of interconnected cavities in their structures. Zeolites are used as molecular sieves to remove water and other small molecules from mixtures, and they can also be employed to separate molecules for which the molecular masses are the same or similar but the molecular structures are different. In addition, they are used as solid supports for highly dispersed catalysts and to promote specific size-dependent chemical reactions.

Silicones

Silicones are polymeric organosilicon compounds containing Si–O–Si linkages and Si–C bonds. They are generally very stable, because of the presence of strong silicon-oxygen and silicon-carbon bonds. A general formula for silicones is $(R_2SiO)x$, where R can be any one of a variety of organic groups. Silicones may be linear, cyclic, or cross-linked polymers, as shown here.

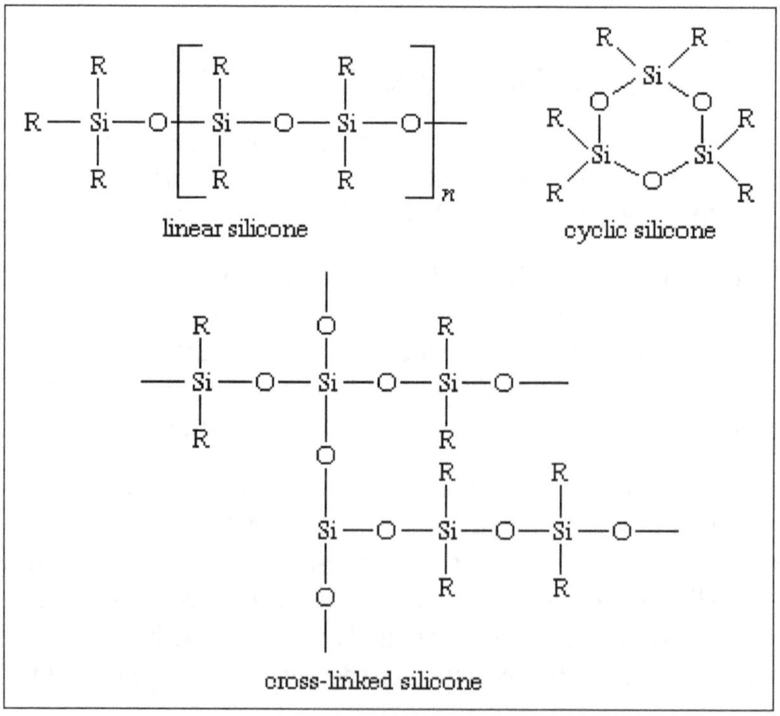

Linear and cyclic silicones are produced by the reaction of water with organochlorosilanes of the general formula R_2SiCl_2, followed by a polymerization reaction that occurs by the elimination of a molecule of water from two hydroxyl groups of adjacent $R_2Si(OH)_2$ molecules.

$$R_2SiCl_2 + 2H_2O \longrightarrow R_2Si(OH)_2 + 2HCl$$

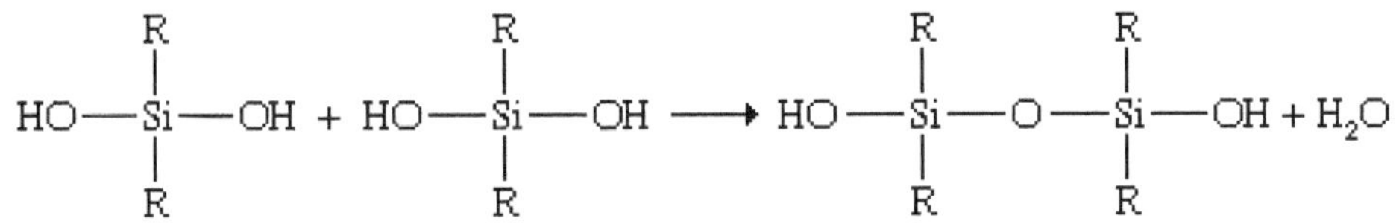

Silicone polymers incorporate some of the properties of both carbon-hydrogen compounds and silicon-oxygen compounds. They are stable to many chemical reagents and to heat. Depending on their degree of polymerization and the complexity of the attached organic groups, silicones can occur in the form of oils, greases, rubberlike substances, or resins. They are used as lubricants, hydraulic fluids, and electrical insulators. They are especially useful as lubricants in applications where there are extreme variations in temperature, because their viscosity changes very little as the temperature changes. Silicones are also water-repellent. Paper, wool, silk, and other fabrics can be coated with a water-repellent film by exposing them for a short time (one to two seconds) to the vapour of trimethylchlorosilane, $(CH_3)_3SiCl$. The $-OH$ groups on the surface of the materials react with the silane, and the surface becomes coated with a thin water-repellent film of $(CH_3)Si-O-$ groups.

$$\text{surface}-OH + Cl-Si(CH_3)_3 \rightarrow (\text{surface}-O-Si(CH_3)_3 + HCl$$

Silanes

Silanes are compounds of silicon and hydrogen. Silicon forms a series of hydrides that have the general formula $\text{Si}n\text{H}_2n_{+2}$, including SiH_4, Si_2H_6, Si_3H_8, and Si_4H_{10}. These compounds contain $Si-H$ and $Si-Si$ single bonds. Silicon possesses empty valence-shell *d* orbitals, which cause the chemistry of silanes to be quite different from that of the corresponding carbon-hydrogen compounds (hydrocarbons). For example, silanes inflame spontaneously in air, whereas the corresponding hydrocarbons do not.

The simplest silane, SiH_4, is called silane. It has a formula and a tetrahedral structure analogous to the hydrocarbon methane, CH_4. It can be prepared in the laboratory by the addition of aqueous acid to an ionic silicide such as magnesium silicide, Mg_2Si.

$$Mg_2Si + 4H_3O^+(\text{aqueous}) \rightarrow 2Mg^{2+} + SiH_4 + 4H_2O$$

Silane is a colourless gas that is thermally stable at normal temperature but reacts violently with air to produce silicon dioxide and water.

$$SiH_4 + 2O_2 \rightarrow SiO_2 + 2H_2O$$

Silanes react readily with hydrogen halides to produce halogenated silanes. For example, silane reacts in a stepwise manner with hydrogen bromide, HBr, to yield all possible combinations of brominated silane from SiH_3Br to $SiBr_4$. Silanes are also very reactive toward hydroxides, producing silicate ions and hydrogen gas.

$$SiH_4 + 2OH^- + H_2O \rightarrow SiO_3^{2-} + 4H_2$$

In contrast, hydrocarbons are inert to hydroxides. The most important reaction of silanes from a commercial standpoint is their reaction with alkenes—i.e., hydrocarbons that contain a carbon-carbon double bond. For example, the reaction of dichlorosilane, H_2SiCl_2, with propene, $H_3CCH{=}CH_2$, yields a diorganodihalosilane that can be used in the preparation of silicones.

$$2H_3CCH{=}CH_2 + H_2SiCl_2 \rightarrow 2(H_3CCH_2CH_2)_2SiCl_2$$

Organic Polymers

Organic polymers are macromolecules composed of many repeating monomer units. Both synthetic and natural polymers play a crucial role in everyday life. Polysaccharides, polypeptides, and polynucleotides are the main types of biopolymers in living cells. These polymers are synthesized by enzyme-mediated processes in cells. In general, synthetic polymers are derived from monomers that contain either a multiple bond, or two or more functional groups, or a three- to seven- membered ring. The chemical properties of the polymers are derived from their monomer units, while the physical properties of polymers are different. Polymers, depending on their physical properties, are characterised as thermoplastics, thermo sets, elastomers and fibers.

Polymers have wide range of applications. In the biomedical filed, biodegradable polymers find immense use in drug delivery and tissue engineering. Neoprene is used in shoe soles, and wet suits. Polymers are extensively used in textiles and fabrics; teflon in non-stick pans and polyvinyl chloride in pipes. The popular PET bottles are made of a synthetic polymer, polyethylene terephthalate. Polystyrene resins are used in the production of home electronics and appliances. Nylon-6 is extensively used in the textile and plastic industries. Poly(methyl methacrylate), a glass-like polymer, is used in a large number of consumer products.

Fluoropolymers

An organic polymer containing fluorine atoms, such as PTFE. A fluoropolymer is a polymer that is based in fluorocarbon and has strong carbon-fluoride bonds. But that doesn't explain much if you don't know what a polymer is. A polymer is a chemical compound that is made of many molecules that have low mass. A polymer is a huge class of compounds; it includes both synthetics, like plastic, and also naturally occurring polymers like wool, silk, and rubber. A fluoropolymer is a more specific type of a polymer that has special properties.

So, more simply put, a fluoropolymer is a chemical compound that contains many strong carbon-fluoride bonds. The fluorine in a fluoropolymer is electrically negative, which gives it the property of not bonding easily with other materials. That's what makes a fluoropolymer non-stick. A fluoropolymer is an umbrella term, and it includes many other products with trade names that you may have heard of, including Teflon, Excalibur, Algoflon, Xylan, Solef, and Fluon. The first fluoropolymers was developed in the 1930s, but hundreds of variations now exist.

Fluoropolymers are in high demand for use in industrial coatings because they possess some very desirable qualities. First, fluoropolymers have excellent non-stick properties, as evidenced by their use in non-stick cookware. Also, fluoropolymers reduce friction and resist corrosion. They withstand very high temperatures, as well. Fluoropolymers are insulators, meaning they do not conduct electricity. They do not absorb water, either. This combination of factors makes them ideal for use in electronics, automotives, and many other uses.

Fluoropolymer coatings protect the products that they are applied to which makes the product last longer and function better through its lifespan. For example:

- Fluoropolymer coatings are applied to the wiring insulation in airplanes: This application protects the wiring and reduces the risk of fire on aircrafts. The next time you are flying across the country, you can rest easy knowing that your plane is protected from fire thanks to fluoropolymer coatings.
- Fluoropolymers are used in many industrial settings: Printing and packaging factories use fluoropolymers to coat their ink trays, rolls, and frames. The coating helps moving parts to not stick to items that pass through, and it also makes them easier to clean.
- Fluoropolymer coatings are also widely used in the food industry: The application of fluoropolymer coatings to molds and trays aids in the production of bread, candy, cheese, and other products by preventing sticking and making the molds and trays easy to clean.
- A big demand for fluoropolymer coatings comes from the automotive industry: Fluoropolymer coatings help to prevent friction and corrosion on car parts. They are particularly useful in components like ball bearings and gears, which must withstand a lot of wear and tear. This application extends the lifespan of your car's parts.

The demand for fluoropolymers is huge; in 2011, it hit $7.25 billion. New uses and applications for fluoropolymers are being developed constantly, and as a result, the demand continues to grow.

Fluoropolymers command a lot of interest because of their extreme chemical resistance. Chemical structures of monomer and polymers, manufacturers and trade names along with typical end-uses of the plastics are included besides extensive tabular and graphical permeation properties. Plastics are the several types of polytetrafluoroethylene (PTFE) and its modifications, fluorinated ethylene propylene (FEP), perfluoroalkoxy (PFA), amorphous fluoropolymer (AF), polyvinyl fluoride (PVF), polychlorotrifluoroethylene (PCTFE), polyvinylidene fluoride (PVDF), ethylene–tetrafluoroethylene (ETFE) copolymer, and ethylene–chlorotrifluoroethylene (ECTFE) copolymer.

Traditionally, a fluoropolymer or fluoroplastic is defined as a polymer consisting of carbon (C) and fluorine (F). Sometimes, these are referred to as perfluoropolymers to distinguish them from partially fluorinated polymers, fluoroelastomers, and other polymers that contain fluorine in their chemical structure. The fundamental properties of fluoropolymers evolve from the atomic structure of fluorine and carbon and their covalent bonding in specific chemical structures. The backbone is formed of carbon–carbon bonds and the pendant groups are carbon–fluorine bonds. Both are extremely strong bonds. Polytetrafluoroethylene polymer (PTFE) is an example of a linear fluoropolymer. The size of the fluorine atom allows the formation of a uniform and continuous covering around the carbon–carbon bonds and protects them from chemical attack, thus imparting chemical resistance and stability to the molecule. Other example of fluoropolymers includes polyethylene chlorotrifluoroethylene, polyethylene tetrafluoroethylene, fluorinated ethylene propylene, perfluoro alkoxy, polychlorotrifluoroethylene, polyvinylidene fluoride, and hexafluoropropylene–tetrafluoroethylene–ethylene copolymer (THE).

The properties of neat or pure fluoropolymers can have a direct impact on fluorocoatings containing them. These are thoroughly documented in other books of the PDL Hand Book Series. Mechanical properties such as tensile strength, modulus, and elongation affect coating durability and postformability (bending and stretching). Thermal properties affect the continuous use temperature of coating systems. Tribology (friction and wear), permeability, weathering, and irradiation properties can also be important.

Polytetrafluoroethylene (PTFE)

Polytetrafluoroethylene or PTFE is the commonly used versatile, high-performance fluoropolymer made up for carbon and fluorine atoms. One of the common applications of this polymer is non-stick coating in kitchen cookware (pans, baking trays etc.), hence, you can easily find this in your kitchen.

Apart from used in the kitchen, PTFE is used as a cost-effective solution for industries ranging from oil & gas, chemical processing, industrial to electrical/electronic and construction sector, etc.

The basic properties of PTFE which make it an interesting material with high commercial value are:

- Exception chemical resistance.
- Good resistance to heat and low temperature.
- Good electrical insulating power in hot and wet environments.
- Good resistance to light, UV and weathering.
- Low coefficient of friction.
- Low dielectric constant/dissipation factor.
- Strong anti-adhesion properties.
- Flexibility.
- Good fatigue resistance under low stress.
- Availability of food, medical and high-purity grades.
- Low water absorption.

PTFE is a linear polymer of tetrafluoroethylene (TFE). It is manufactured by free-radical polymerization mechanism in an aqueous media via addition polymerization of TFE in a batch process.

The chemical structure of PTFE $[CF_2\text{-}CF_2]_n$ is like that of polyethylene (PE), except that the hydrogen atoms are completely replaced by fluorine (hence it is referred as perfluoro polymer). However, it is important to note that in practice PTFE and PE are prepared and used in totally different ways.

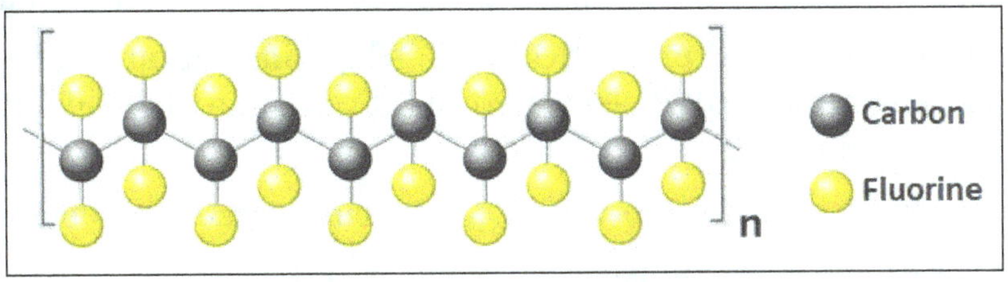

Molecular Structure of PTFE.

It is the size of fluorine atom which forms a uniform and continuous sheath around carbon-carbon-bonds and hence imparts good chemical resistance and stability to the molecule. This uniform fluorine sheath also provides electrical inertness to the molecule. The fluorine content in PTFE is theoretically 76% and it has 95% crystallinity.

PTFE was first discovered "accidentally" in 1938 by Dr. Plunkett at DuPont. After that PTFE was made commercially available in 1947 with the trademark "Teflon™" from Chemours. It was the discovery of PTFE that accelerated the development of the other fluoropolymers.

Typical Characteristics and Properties of PTFE

PTFE is available in granular, fine powder and water-based dispersion forms:

- The granular PTFE resin is produced by suspension polymerization in an aqueous medium with little or no dispersing agent. Granular PTFE resins are mainly used for molding (compression and isostatic) and ram extrusion.
- The fine PTFE powder is prepared by controlled emulsion polymerization, and the products are white, small sized particles. Fine PTFE powders can be processed into thin sections by paste extrusion or used as additives to increase wear resistance or frictional property of other materials.
- PTFE dispersions are prepared by the aqueous polymerization using more dispersing agent with agitation. Dispersions are used for coatings and film casting.
- PTFE has excellent properties such as chemical inertness, heat resistance (both high and low), electrical insulation properties, low coefficient of friction (static 0.08 and dynamic 0.01), and nonstick property over a wide temperature range (260 to 260 °C).
- It has a density in the range of 2.1 - 2.3 g/cm^3 and melt viscosity in the range of 1 -10 GPa per second.
- PTFE is among the most chemically resistant polymer. The exceptions include molten alkali metals, gaseous fluorine at high temperatures and pressures, and few organic halogenated compounds such as chlorine trifluoride (ClF_3) and oxygen difluoride (OF_2). View PTFE Grades With Good Chemical Resistance.

- Mechanical properties of PTFE are generally inferior to engineering plastics at the room temperature. Compounding with fillers has been the strategy to overcome this shortage. PTFE has useful mechanical properties in its use temperature range.

  The mechanical properties of PTFE are also affected by processing variables such a preform pressure, sintering temperature, cooling rate etc. Polymer variable such as molar mass, particle size, particle size distribution poses significant impact on mechanical properties.
- PTFE has excellent electrical properties such as high insulation resistance, low dielectric constant. has an extremely low dielectric constant (2.0) due to the highly symmetric structure of the macromolecules.
- PTFE exhibits high thermal stability without obvious degradation below 440 °C.
- PTFE materials can be continuously used below 260 °C.
- PTFE is attacked by radiation, and degradation in air begins at a dose of 0.02 Mrad.

These properties come from the special electronic structure of the fluorine atom, the stable carbon-fluorine covalent bonding, and the unique intramolecular and intermolecular interactions between the fluorinated polymer segments and the main chains.

Property	Value
Melting Temperature (°C)	317-337
Tensile Modulus (MPa)	550
Elongation at Break (%)	300-550
Dielectric strength (kV/mm)	19.7
Dielectric Constant	2.0
Dynamic Co-efficient of Friction	0.04
Surface Energy (Dynes/g)	18
Appl. Temperature (°C)	260
Refractive Index	1.35

Limitation of PTFE

The conventional PTFE has some limitations in its applications, such as:

- Impossibility of using conventional molten-state processing methods and difficulty and cost of the suitable specific methods.
- Sensitivity to creep and abrasion.
- Significant dimensional variation around glass transition temperature (19 °C).
- Difficulties of joining.
- Corrosive and prone to toxic fumes.
- Low radiation resistance.

Impact of Fillers and Additives on PTFE Properties

Mechanical properties of PTFE can be enhanced with the addition of fillers, particularly creep and wear rate. Glass fiber, bronze, steel, carbon, carbon fiber, graphite, etc. are among the common fillers used.

Glass fiber has a positive impact on creep performance of PTFE by reducing its low and high temperatures. Glass filled compounds perform well in oxidizing environments. Further, PTFE's wear characteristics are improved.

Carbon reduces creep, increases hardness and elevates thermal conductivity of PTFE. When combined with graphite, the wear resistance of carbon-filled compounds can be improved further. These compounds are well suited for non-lubricated applications such as piston rings in compressor cylinder. Further, graphite imparts excellent wear properties to PTFE and graphite-filled PTFE has an extremely low coefficient of friction.

Carbon fiber lowers creep, increases flex and compressive modulus and raises hardness. Unlike glass fibers, carbon fibers are inert to hydrofluoric acid and strong bases. Carbon fiber PTFE compounds have lower coefficient of thermal expansion and high thermal conductivity. These parts are ideal for automotive parts in shock absorbers, water pumps etc.

Bronze-filled PTFE compounds have high thermal and electrical conductivity which is in turn make these compounds well fit for application where a part is subjected to load in extreme temperatures.

Other fillers which are incorporated in PTFE to produce specialty compounds include Calcium fluoride, Alumina, Mica, polymeric fillers.

In general:

- Fillers result in excellent properties of PTFE at low and high temperatures.
- Fillers/additives increase the porosity of PTFE compounds and hence impact electrical properties – dielectric strength decreases while dielectric constant and dissipation factor increases.
- Chemical properties well depend on type of filler used. In general, chemical properties of filled PTFE compounds are not as good as those of unfilled resin.
- Filler impart change in electrical and thermal conductivity of PTFE.

Up to 40% by volume of filler can be added to the PTFE without complete loss of physical properties. The impact of fillers below 5% is low.

Applications of Polytetrafluoroethylene (PTFE)

Fluorinated thermoplastics are only used for high performance applications related to high heat, low temperature, chemical inertness, high purity, non-sticking and self-lubricating properties. Further high purity grades are used in semiconductor, pharmaceutical and other similar sectors.

Due to its special chemical and physical properties, PTFE are widely applied in the chemical, electrical/electronic, construction, architectural, and automotive industries.

Applications of Polytetrafluoroethylene (PTFE)

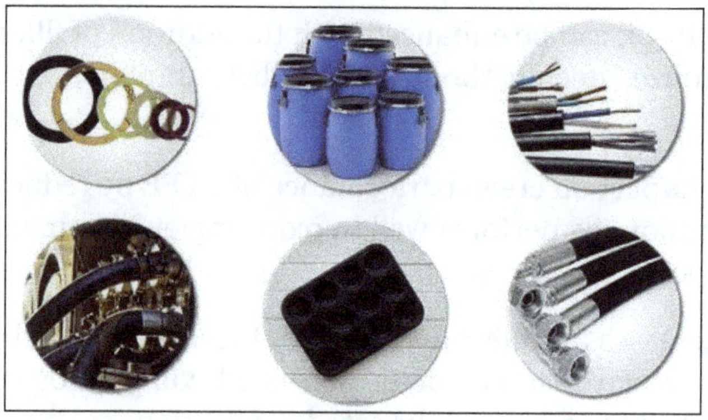

Filled granular resins were found to be suitable for parts such as gaskets, shaft seals, bearing, piston rings etc.

Automotive	O-rings, gaskets, valve stem seals, shaft seals, linings for fuel hoses, power steering and transmission etc.
Chemical Industry	Coatings for heat exchangers, pumps, diaphragms, impellers, tanks, reaction vessels, autoclaves, containers etc.
Electrical & Electronics	Electrical insulation, flexible printed circuit boards, semiconductor parts.
Engineering	Seats and plugs, bearing, non-stick surfaces, coatings for pipes, fittings, valve and pump parts.
Medical	Cardiovascular grafts, ligament replacement, heart patches.

Popular Techniques used to Process PTFE

PTFE has the very high melt viscosity and a high melting temperature due to rigid polymer chain structure that makes processing difficult by the normal methods of extrusion and injection molding. Processing technologies have more similarity to those of powder metallurgy than those of traditional plastics processing.

- Sintering, pressing, ram or paste extrusion, compression molding or isotactic molding, machining, hot stamping and extrusion of presintered powders on special machines.
- Paste extrusion in which PTFE is blended with a hydrocarbon, prior to molding a preform, is used to continuously fabricate PTFE into tubes, tapes, and wire insulation. The hydrocarbon is vaporized before the part is sintered.
- Dispersion – metal coatings, coatings, pulverization, impregnation, cast for thin films and fiber spinning.
- Operating range (-270 °C) -200 °C to 260 °C (280 °C).

The properties of the PTFE products are strongly dependent on the processing procedure, such as polymer particle size, sintering temperature and processing pressure. Therefore, other fluoropolymers are still needed for some specific applications where PTFE is not completely suitable.

This led to a search for melt-processable fluoropolymers and the development of other members of the family.

References

- Inorganic-polymers: britannica.com, Retrieved 9 May, 2019
- Organic-polymers: alfa.com, Retrieved 31 March, 2019
- What-is-fluoropolymer-coating: toefco.com, Retrieved 14 July, 2019
- Fluoropolymer, materials-science: sciencedirect.com, Retrieved 17 May, 2019
- Polytetrafluoroethylene-ptfe-fluoropolymer, selection-guide: specialchem.com, Retrieved 5 February, 2019

4

Polymerization Reactions

Polymerization is the reaction between molecules in a chemical reaction in order to form three-dimensional network or polymer chains. The major types of polymerization are coordination polymerization, dispersion polymerization, step-growth polymerization, nitroxide-mediated radical polymerization, ring-opening polymerization, etc. All these types of polymerization have been carefully analyzed in this chapter.

Polymerization

Polymerization is a process in which relatively small molecules, called monomers, combine chemically to produce a very large chainlike or network molecule, called a polymer. The monomer molecules may be all alike, or they may represent two, three, or more different compounds. Usually at least 100 monomer molecules must be combined to make a product that has certain unique physical properties—such as elasticity, high tensile strength, or the ability to form fibres—that differentiate polymers from substances composed of smaller and simpler molecules; often, many thousands of monomer units are incorporated in a single molecule of a polymer. The formation of stable covalent chemical bonds between the monomers sets polymerization apart from other processes, such as crystallization, in which large numbers of molecules aggregate under the influence of weak intermolecular forces.

Two classes of polymerization usually are distinguished. In condensation polymerization, each step of the process is accompanied by the formation of a molecule of some simple compound, often water. In addition polymerization, monomers react to form a polymer without the formation of by-products. Addition polymerizations usually are carried out in the presence of catalysts, which in certain cases exert control over structural details that have important effects on the properties of the polymer.

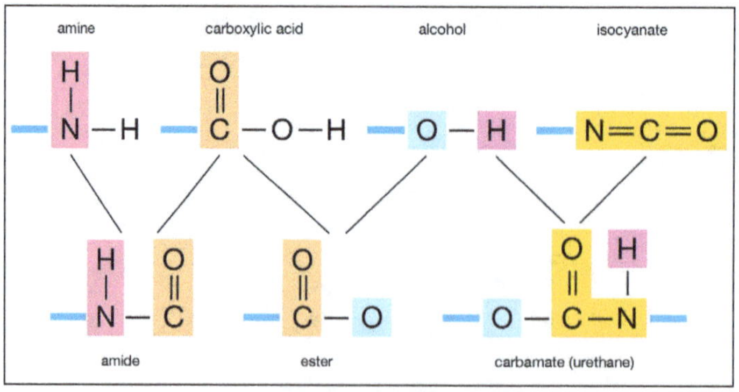

Functional group in monomers and polymers.

Linear polymers, which are composed of chainlike molecules, may be viscous liquids or solids with varying degrees of crystallinity; a number of them can be dissolved in certain liquids, and they soften or melt upon heating. Cross-linked polymers, in which the molecular structure is a network, are thermosetting resins (i.e., they form under the influence of heat but, once formed, do not melt or soften upon reheating) that do not dissolve in solvents. Both linear and cross-linked polymers can be made by either addition or condensation polymerization.

Degree of Polymerization

Degree of polymerization is an average number of monomers (mers) in a polymer molecule.

Step-growth/Condensation Polymerization

Polycondensation Reactions

Polycondensation reactions are those chemical reactions in which the polymer is originated by successive bonds between monomers, which emit condensed molecules during the bonding process.

Condensed molecules that are released to the environment, due to the polycondensation process depends on the nature of the monomers that will bond in order to made the polymer, for example, silicone-based adhesives of 2 components that are produced during the polycondensation reaction curing stage emitted alcohols to the environment, as the result of the condensed molecules produced during the polycondensation reaction.

Condensed molecules that arise during the polycondensation process are molecules with low molecular weight such as water, hydrogen chloride, alcohols, etc., which are in gaseous state, separating from the resulting polymer via evaporation.

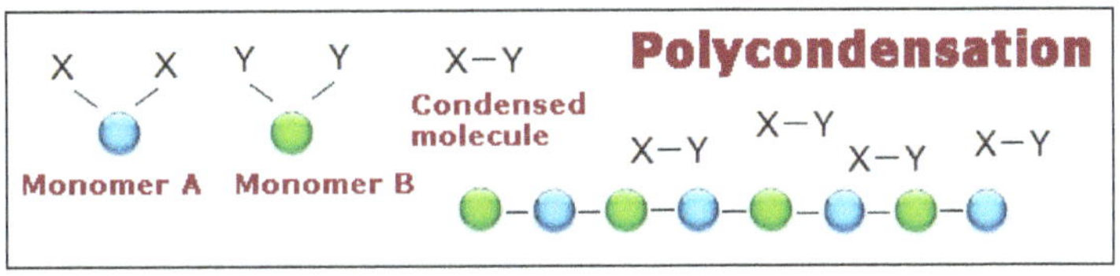

Curing adhesives by polycondensation:

- Silicones.
    - Adhesives 2-component silicone.
    - Silicone Adhesives 1 component (moisture).
- Modified Silane Polymer.
- Polyamides.
- Polysulfide.
- Phenol resins.
- In general the rate of cure of the adhesive by polycondensation is less than other adhesives.

Copolymerization

Copolymerization involves 2 monomers so that the propagation involves 4 possibilities of steps.

$$A \quad \sim M_1^\bullet + M_1 \xrightarrow{k_{11}} \sim M_1M_1\bullet$$
$$B \quad \sim M_1^\bullet + M_2 \xrightarrow{k_{12}} \sim M_1M_1\bullet$$
$$C \quad \sim M_2^\bullet + M_1 \xrightarrow{k_{21}} \sim M_2M_1\bullet$$
$$D \quad \sim M_2^\bullet + M_1 \xrightarrow{k_{22}} \sim M_2M_2\bullet$$

The rate equations are:

$$\frac{-d[M_1]}{dt} = k_{11}\left[\sim M_1^\bullet\right][M_1] + k_{21}\left[\sim M_2^\bullet\right][M_1]$$
$$\frac{-d[M_2]}{dt} = k_{12}\left[\sim M_1^\bullet\right][M_2] + k_{22}\left[\sim M_2^\bullet\right][M_2]$$

The copolymer composition depends on the comonomer reactivity and concentration. This means that the monomer consumed during copolymerization depends on the polymer structure.

The instantaneous copolymer equation describes the relative change in the monomer composition.

$$\frac{d[M_1]}{d[M_1]} = \frac{[M_1]}{[M_2]} \frac{k_{11}\left[\sim M_1^\bullet\right] + k_{21}\left[\sim M_2^\bullet\right]}{k_{12}\left[\sim M_1^\bullet\right] + k_{22}\left[\sim M_1^\bullet\right]}.$$

Model I

The reactivity ratio: $r_1 = \frac{k_{11}}{k_{12}} \quad r_2 = \frac{k_{22}}{k_{21}}$

Explanation:

r_1 is the reactivity ratio for monomer 1, = the ratio of propagation rate-constants for the addition of M_1 (homopol' n) and addition of M_2 (copol' n) to a propagating radicalchain with monomer 1 at the radical end ($\wedge\vee\wedge\vee M_1\bullet$).

r_2 is the reactivity ratio for monomer 2, = the ratio of propagation rate-constants for the addition of M_2 (homopol' n) and addition of M_1 (copol' n) to a propagating radicalchain with monomer 2 at the radical end ($\wedge\vee\wedge\vee M_2\bullet$).

Some cases:

- Both reactivity ratios are equal ($k_{11} = k_{12}$ and $k_{22} = k_{21}$), there is no preferential monomer incorporation into the propagating chain or the copolymer is random. (Example: styrene and 4-chlorostyrene).
- Both reactivity ratios are zero ($k_{11} = k_{22} = 0$), alternating copolymers (Example: styrene and maleic anhydride).

Some cases:

- Both reactivity ratios are small but not exactly zero ($k_{11} = k_{22} \neq 0$ and <<<), the copolymer are highly alternating with some random placement. (Example: styrene and acrylonitrite).
- Both reactivity are very much larger than unity (kii >>> kjj), the copolymers are homopolymers only or block copolymers only.

The limitation of the instantaneous copolymer equation is that the radicalchain concentrations is usually not known or easily measured. Assume that a steady-state radical-concentration for either (∧∨∧∨ M_1•) and (∧∨∧∨ M_2•) is reached during propagation. Mayo and Lewis used the definitions of reactivity ratios and the i.c.equation becomes:

$$\frac{d[M_1]}{d[M_2]} = \frac{[M_1]}{[M_2]}\left(\frac{r_1[M_1]+[M_2]}{[M_2]+r_2[M_2]}\right)$$

For a given monomer concentration, the Mayo-Lewis i.c equation can be used to calculate the copolymer composition. It is defined that:

F_1 = the mole fraction of monomer 1 in the monomer mixture and $f_1 + f_2 = 1$,

F_1 = the mole fraction of monomer 1 in the copolymer and $F_1 + F_2 = 1$.

$$f_1 = \frac{[M]}{[M_1]+[M_2]} \qquad F_1 = \frac{d[M_1]}{d[M_1]+d[M_2]}$$

$$F_1 = \frac{r_1 f_1^2 + f_1 f_2}{r_1 f_1^2 + 2 f_1 f_2 + r_2 f_2^2}$$

Some conditions:

- If $r_1 = r_2 = 1$ then $F_1 = f_1$, ideal of statistical copolymerization (---).
- If $r_1 = r_2 = 0$ then $F_1 = 0.5$, alternating polymer (– – –).
- If r_1 and r_2 are random, partially alternating (———).

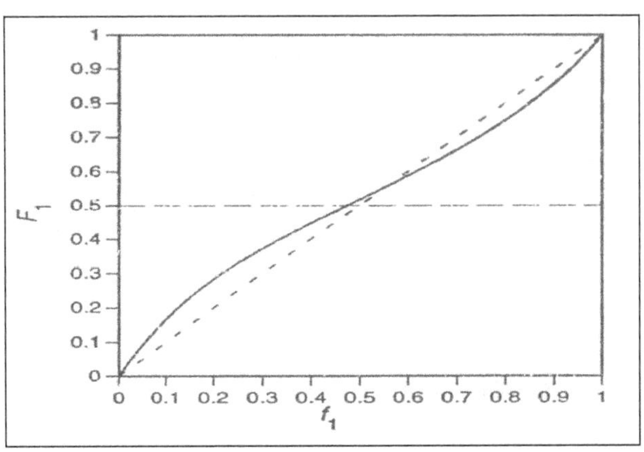

Plot of mole fraction of monomer 1 in copolymer F_1, versus mole fraction of monomer 1 in the feed f_1, for (---) ideal ($r_1 = r_2 = 1$), (---) alternating ($r_1 = r_2 = 0$), and (---) partially alternating copolymerization (copolymerization of styrene and methyl methacrylate, where $r_1 = 0.585$ and $r_2 = 0.478$).

Example: Partially alternating copolymers of styrene and methyl methacrylate, $r_1 = 0.585$ and $r_2 = 0.478$.

In fact, neither ideal (random) of perfectly alternating happens. Most of reactive monomers will be preferentially consumed.

The monomer feed composition varies with time.

- t >>>, high monomer conversion → heterogeneous mixture of individual copolymers having different composition → low mechanical strength.
- t <<<, low monomer conversion → high cost copolymer recovery.

Table: Example of reactivity rations for free radical copolymerization.

Monomer 1	Monomer 2	R_1	R_2
Ethylene	Vinyl acetate	0.130	1.230
	Carbon monoxide	0.025	0.004
	Propylene	3.200	0.620
Styrene	Acrylonitrile	0.290	0.020
	Butadiene	0.820	1.380
	Divinylbenzene	0.260	1.180
	Maleic anhydride	0.097	0.001
	Methyl methacrylate	0.585	0.478
	4-Chlorostyrene	0.816	1.062
	Vinylidene chloride	1.700	0.110
Vinyl chloride	Vinylidene chloride	0.205	3.068

The value of a reactivity ratio is independent of the nature of the initiator and the solvent in a free-radical copol'n, but there is a weak temperature dependence.

Model II (Alfrey and Price)

The propagation rate-constant, k_{ij} is defined as:

$$K_{ij} = P_i Q_i \exp\left(-e_i e_j\right)$$

Where:

- P_i is a proportionality constant,
- Q_i is a measure of the monomer reactivity,
- e_1 and e_2 are the polarity of the radical $M_1\bullet$ and $M_2\bullet$ respectively.

The reactivity ratio:

$$r_1 = \frac{k_{11}}{k_{12}}\left(\frac{Q_1}{Q_2}\right)\exp\left[-e(e_1 - e_2)\right]$$

$$r_2 = \frac{k_{22}}{k_{21}}\left(\frac{Q_2}{Q_1}\right)\exp\left[-e(e_2 - e_1)\right]$$

Table: Q-e values for free-radical copolymerzation.

Monomer	Q	e
Acrylamide	0.23	0.54
Acrylonitrile	0.48	1.23
Butadiene	1.70	-0.50
Ethylene	0.016	0.05
Isobutylene	0.023	-1.20
Isoprene	1.99	-0.55
Maleic anhydride	0.86	3.69
Methacrylic acid	0.98	0.62
Methyl methacrylate	0.78	0.40
N-Vinyl pyrrolidone	0.088	-1.62
Styrene	1.00	-0.80
Vinyl acetate	0.026	-0.88
Vinyl chloride	0.056	0.16
Vinylidene chloride	0.31	0.34

Extensions of Copolymerization Kinetics

Multicomponent Copolymerization

The copolymer equation for 3 or more monomers are quite complex. However, some simple approaches, such as steady-state condition and the prob of finding certain monomers, result in a good agreement with experiment.

Penultimate and Remote-unit Effects

The mer units before the last polymer may affect the reactivity of chain radical.

Depropagation during Copolymerization

A monomer or both monomers has/have a tendency to depropagate during copolymerization. This deviation can be observed by the dependence of the copolymer composition on the absolute concentration of the monomers (and temperature), even though their concentration vary.

Complex Participation

The presence of complexes between comonomers can cause deviation from normal copolymerization kinetics. This behaviour can be observed from experimental data.

Composition of Copolymers

$$\frac{d[M_1]}{d[M_2]}=\frac{[M_1]}{[M_2]}\left(\frac{r_1[M_1]+[M_2]}{[M_1]+r_2[M_2]}\right)$$

This copolymer equation can predict the average composition of the polymer formed at any instant in the polymerization. There 3 methods to determine the copolymer composition.

Statistical Fluctuation of i.c. Composition

Alternating copolymers give the minimum spread in composition.

$$N(n)=p^{n-1}(1-p) \qquad p=\frac{r_1[M_1]}{r_1[M_1]+[M_2]}$$

Where N is the prob to find n similar monomer unit in an ideal copolymer, p is the prob of addition of that monomer

Integration of Copolymer Equation

$$F_1=\frac{r_1 f_1^2+f_1 f_2}{r_1 f_1^2+2f_1 f_2+r_2 f_2^2}$$

For a system in which $F_1 > f_1$, when dM moles of monomer have polymerized, the polymer contains F_1 dM moles of monomer 1, and the feed contains $(f_1 - df_1)$ (M - dM) moles of monomer 1.

For a material balance:

$$M\,f_1-(M-dM)(f_1-df_1)=F_1\,dM \rightarrow \frac{dM}{M}=\frac{df_1}{F_1-f_1}$$

$$\varepsilon=\frac{1-r_2}{(2-r_1+r_2)}$$

$$\log\left(\frac{M}{M_0}\right)=\frac{r_2}{1-r_2}\log\left(\frac{f_1}{(f_1)_0}\right)+\frac{r}{1-r_1}\log\left(\frac{f_2}{(f_2)_0}\right)+\frac{1-r_1r_2}{(1-r_1)(1-r_2)}\log\left(\frac{(f_1)_0-\varepsilon}{f_1-\varepsilon}\right)$$

Variation of Copolymer Composition with Conversion

An ideal system: styrene – vinylthiophene, r_1 = 0.35 and r_2 = 3.10. The greater reactivity of the vinylthiophene causes it to be depleted until at the last only styrene remains, and pure PS is the final product.

An alternating system: styrene – diethyl fumarate, $r_1 = 0.30$ and $r_2 = 0.07$. A strong tendency to alternate and an azeotropic composition at 57 mol% styrene.

Homogeneity in Copolymer

- Styrene – diethyl fumarate copolymer containing 40% styrene, stopped at 75% conversion, contains only 44 – 52% styrene.
- Feed the desired polymer and add more reactive monomer.

Mechanisms of Copolymerization

Free-radical Copolymerization

The behaviour of monomers in copolymerization is determined by 2= factors: the general monomer reactivity and the alternating tendency. These depend on the nature of the attacking radical.

The reactivity of monomers and radicals is determined by the nature of the substituents on the double bond of the monomer.

There are 3 ways that these substituents influence the reactivity. They may:

- Activate the double bond, making the monomer more reactive,
- Stabilize the resulting radical by resonance,
- Provide steric hindrance at the reaction site.

Reactivity of Monomers

The relative reactivities of monomers to a reference radical can be derived from the monomer reactivity ratio. The inverse of this ratio is the rate of reaction of the reference radical with another monomer.

Resonance Stabilization

Resonance stabilization of monomers and radicals can be considered in terms of their potential energy.

Reactivity of Radicals

The reactivity of radicals to a reference monomer can be derived by combining the absolute propagation rate constant (k_{11}) and reactivity ratio to obtain the rate constants for radical 1 adding monomer 2 (k_{12}).

Steric Effects

The effect of steric hindrance in reducing reactivity may be demonstrated by comparing the reactivities of monomers to reference radicals.

Charge-transfer Complexes

A charge-transfer complex can be formed from 2 monomers with different polarities (e.g. an electron donor and an electron acceptor). This complex may act as a single species in polymerization, leading to 1 : 1 alternating polymers. The formation of such complexes is detected by spectroscopy evidence. Example of electron acceptors: methyl methacrylate, methyl acrylate.

Example of electron donors: vinyl chloride, vinyl acetate, propylene.

Polarity and Alternation

By giving the product of $r_1 r_2$, it is possible to calculate 2 monomers with a tendency to alternate.

The Alfrey-price Treatment

The product of $r_1 r_2$ possibly calculate 2 monomers with a tendency to alternate.

$$r_1 r_2 = \exp\left[-(e_1 - e_2)^2\right]$$

Block and Graft Copolymers

Polymers with long sequences of 2 monomers can have 2 arrangements of chains: the block copolymers and the graft copolymers. The aims of obtaining these types of polymer is having copolymer free from homopolymers to enhance physical properties.

Block Copolymers

The 2 monomers follow one another along the main polymer chain.

–AABB – BBAA – AABB –

The usual method: Using a polymer with end groups that can be made to react under different conditions with different ways.

- Labile end groups: Labile end groups can be activated the thermal or uv scission of the weak bond.
- Disfunctional initiators: Two initiator groups can be decomposed independently.
- Living polymers: Unterminated anionic polymerization produces "living" polymers.

Graft Copolymers

One monomer are "grafted" onto a "backbone" the 2nd monomer.

```
–AAA – AAA – AAA –
  B     B     B
  B     B     B
  |     |     |
```

Graft copolymer is a result from the formation of an active site at a point on a polymer molecule and exposure to a 2nd monomer. Most graft copolymer are formed by radical polymerization after uv or ionizing radiation, or redox initiation.

Ideal Copolymerization (Ideal Random Vinyl Copolymerization)

In the ideal random vinyl copolymerization, each radical bearing chain has no preference for adding monomers of type M_1 or M_2. Furthermore, both types of radical bearing end groups have the same rate of addition. This means, the probability of the addition of a M_1 monomer to a $M_2\cdot$ radical is the same as the addition of a M_2 monomer to a $M_2\cdot$ radical. The same is true for the addition of monomers to a $M_1\cdot$ radicals. Therefore, the sequence of repeat units in an ideal copolymer is completely random. These polymers are called random or statistical copolymers.

If r_1 and r_2 are monomer reactivity ratios defined by,

$$r_1 = k_{11} / k_{12}$$

$$r_2 = k_{22} / k_{21}$$

then the condition for random copolymerization is,

$$r_1 = 1 / r_2$$

$$k_{21} / k_{22} = k_{11} / k_{12}$$

where k_{xy} are the rate constants of the four different types of free radical polymerization reactions.

$$M_1\cdot + M_1 \rightarrow M_1\cdot \quad R_{11} = k_{11}[M_1\cdot][M_1]$$

$$M_1\cdot + M_2 \rightarrow M_2\cdot \quad R_{12} = k_{12}[M_1\cdot][M_2]$$

$$M_2\cdot + M_2 \rightarrow M_2\cdot \quad R_{22} = k_{22}[M_2\cdot][M_2]$$

$$M_2\cdot + M_1 \rightarrow M_1\cdot \quad R_{21} = k_{21}[M_2\cdot][M_1]$$

Where $M_1\cdot$ and $M_2\cdot$ represent chain radicals having a free radical bearing unit of type 1 and 2. We have to distinguish between two cases:

Case: $r_1 > 1$ *and* $r_2 < 1$ or $r_1 < 1$ *and* $r_2 > 1$, $r_1r_2 = 1$

Each radical has no preference and the copolymer has a completely random sequence of monomers. However, one monomer is more reactive than the other. Consequently, the copolymer will consist of a greater portion of the more reactive specimen in the random sequence of repeat units.

Case: $r_1 = r_2 = 1$

The sequence is not only random, but the composition is identical with the composition of the monomer blend throughout the range. However, the reactivity of the two radicals might differ, that is, k_{11} need not be equal to k_{22}. Incremental polymer composition as a function of monomer composition for ideal copolymerization.

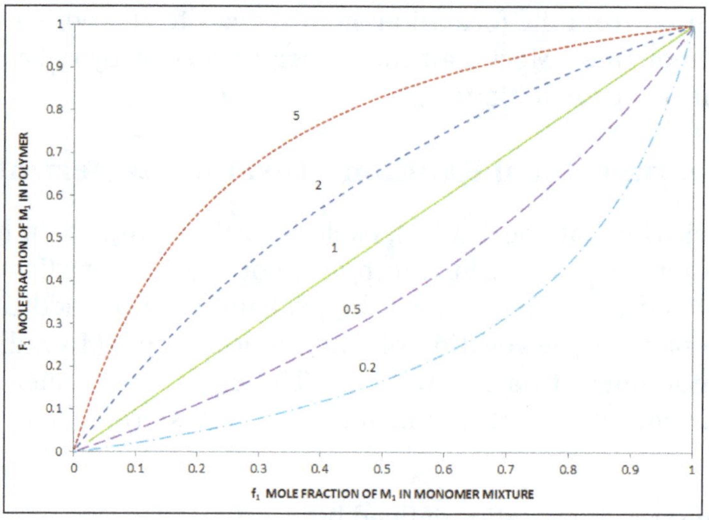

The figure above shows the incremental polymer composition as a function of monomer composition for different reactivity ratios r_1/r_2 of an ideal copolymerization. The ordinate (F_1) represents the composition of the copolymer increment formed from the monomer blend of composition f_1 shown on the abscissa. The straight line $r_1 = 1$ is the trivial case in which,

$$r_1 = r_2 = 1, \Rightarrow k_{12} = k_{11} \text{ and } k_{21} = k_{22}.$$

This means the two radicals are equally reactive towards each monomer. The radicals sequence is then not only random, but the incremental composition is identical with the mole fractions of the unreacted monomer blend.

Alternating Copolymerization

An alternating copolymer consists of two different repeat units that alternate in sequence in the polymer chain. A regularity in repetition of structural units is only possible if the two monomers M_1 or M_2 are unable to hompolymerize or have a strong preference to react with each other (quasi alternating copolymer). This is only the case if the reactivity ratios defined by,

$$r_1 = k_{11} / k_{12}$$

$$r_2 = k_{22} / k_{21}$$

approach zero,

$$r_1 = r_2 \approx 0,$$

Where k_{xy} are the reaction constants of the four different types of polymerization reactions:

$$M_1^* + M_1 \neq> M_1M_1^* \quad R_{11} = k_{11} [M_1^*] [M_1]$$

$$M_1^* + M_2 => M_2M_1^* \quad R_{12} = k_{12} [M_1^*] [M_2]$$

$$M_2^* + M_2 \neq> M_2M_2^* \quad R_{22} = k_{22} [M_2^*] [M_2]$$

$$M_2^* + M_1 => M_1M_2^* \quad R_{21} = k_{21} [M_2^* \cdot] [M_1]$$

M_1^* and M_2^* represent chains terminating in a unit of type 1 and 2. In the case of a free radical polymerization, the radical bearing end groups react only with monomers of the other kind, i.e. $M_1\cdot$ with M_2 and $M_2\cdot$ with M_1.

For a quasi alternating copolymerization, the reactivity ratios are only approxiamtely but not exactly equal and zero. For obvious reasons, in a quasi alternating copolymerization of a 50/50 monomer blend, the two types of monomers will not be consumed with the same rate. Thus, with increasing degree of conversion, the monomer ratio $[M_1]/[M_2]$ will deviate more and more from unity and the polymer structure will become less and less alternating.

Assuming chain termination and transfer reactions are absent or rare events, polymerization continues until one of the monomers is used up and then either stops ($r_1 \approx r_2 \approx 0$) or continues until both monomers are consumed ($r_1 \approx r_2 > 0$). Incremental polymer composition as a function of monomer composition for quasi alternating copolymerization.

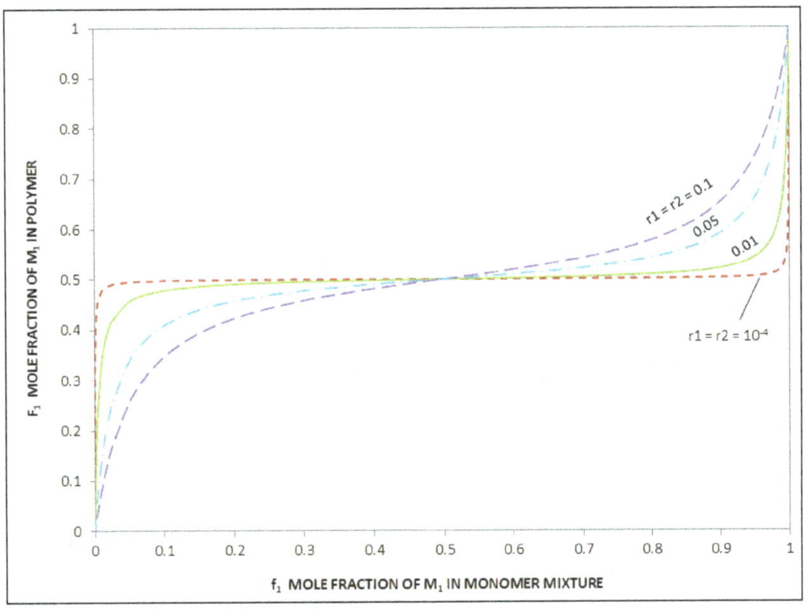

The figure above shows the incremental polymer composition as a function of monomer composition for different reactivity ratios. As the reactivity ratios r_1 and r_2 approach zero, there is an increasing tendency towards regularity in repetition of structural units.

Block Copolymerization

Block copolymers consist of two or more strands (blocks) of different polymers chemically attached to each other. The properties of these polymers depend on the copolymer sequence distribution, the chemical nature of the blocks, the average molecular weight and the molecular weight distribution of the blocks and the copolymer.

Block copolymers, particularly those of type A-B-A can exhibit properties that are very different from those of random copolymers.

In the case of copolymerization of vinyl polymers, block copolymers are often prepared by controlled sequential monomer addition. In this case, one of the two monomers is polymerized first.

When polymerization of the first component is complete, a second (suitable) monomer is added to the "living" radical bearing polymers which then copolymerizes with the first component.

Another method involves monomers that have a strong tendency to homopolymerize but still copolymerize. This is the case, if the reactivity ratios, defined by,

$$r_1 = k_{11} / k_{12}$$

$$r_2 = k_{22} / k_{21}$$

are (much) larger than unity:

$$r_1 \approx r_2 > 1,$$

or

$$k_{11} > k_{12},\ k_{22} > k_{21}$$

where k_{xy} are the reaction constants of the four different types of polymerization reactions:

$$M_1^* + M_1 \rightarrow M_1M_1^* \quad R_{11} = k_{11} [M_1^*] [M_1]$$

$$M_1^* + M_2 \rightarrow M_2M_1^* \quad R_{12} = k_{12} [M_1^*] [M_2]$$

$$M_2^* + M_2 \rightarrow M_2M_2^* \quad R_{22} = k_{22} [M_2^*] [M_2]$$

$$M_2^* + M_1 \rightarrow M_1M_2^* \quad R_{21} = k_{21} [M_2^* \cdot] [M_1]$$

M_1^* and M_2^* represent reactive chain ends of type 1 and 2.

For $r_1 > 1$ and $r_2 > 1$, each radical has a strong preference to add monomers of its kind to the growing polymer chain. The addition of the same type of monomer will continue until there is a chance of addition of the other kind of monomer, and then polymerization of the other monomer will continue. In the case of a very low probability of crosspolymerization ($r >>1$), large M_1 and M_2 blocks will form and in the extreme case of both reactivity ratios very large, $r_1 \approx r_2 \rightarrow \infty$, the two monomers only react with themselves, leading to a mixture of two homopolymers. However, complete free radical homopolymerization in the presence of another monomer has not been observed yet.

Ring-opening Polymerization

Ring-opening polymerization (ROP) is, together with chain (radical and ionic) polymerization and condensation polymerization, one of the three paths to the polymers that are so important to life in the 21st century. Some ROP can be considered as a chain polymerization (addition of monomer to a growing chain end) but many reactions are more complicated and involve, e.g., activated monomers. Particularly ROP has proved to be a useful synthetic route to technologically interesting polymers with very specific, and controllable properties (e.g., refractive index), for preparing synthetic variants to naturally occurring polymers (e.g., chitin,) or to optimize biodegradable polymers for agricultural, medicinal and pharmaceutical applications.

With some exceptions, such as the ROP of Leuchs' anhydridesor the radical ROP of 2,2-diphenyl-4-methylene-1,3-dioxolane, is not accompanied by the elimination of small molecules. Additionally, ROP is not aided by the enthalpy difference between a single and double C–C bond, which counteracts the loss of entropy occurring as a consequence of polymerization via chain polymerization. ROP have in common that the monomers are rings, irrespective of ring size, but the reason why polymerization takes place can vary. Rings composed of 3–8 atoms may polymerize due to the loss of enthalpy associated with the loss of ring strain. Thus, the ring strain for oxiranes is 116 kJ mol^{-1} and even for the 7 and 8 membered lactones and lactams the polymerization is driven by an enthalpic contribution even though, in these cases the ring strain is only *ca.* 6 J mol^{-1}. Strain-free six-membered rings do not generally polymerize. For rings containing disulfide, silicon or carbonate moieties an additional effect is the increased rotational freedom of these groups in the resulting linear chains which leads to an increase in entropy to give the driving force for polymerization.

Many polymers of industrial importance are produced via ROP, e.g., olycyclooctene, polynorbornene, polyethylene oxide, polysiloxane and polyphosphazene. A large family of commercially available polymers called polyethylenimines (many of which are highly branched), are produced via a CROP of aziridine or oxazoline monomers. Polyoxymethylene (POM), derived from trioxane and most of the diols (produced, e.g., via CROP of THF) used in polyurethanes as well as Nylon 6 (poly ε-caprolactam) are perhaps, in terms of volume, the most significant. More recently, polyglycolides and polylactides are becoming important as packaging materials due to the bio-compatibility and natural decomposition of these materials.

Another important development is the use of metathesis catalysts to open rings with exo functional groups to yield novel polymer structures. Table lists the major groups of ring structures and the mechanism by which they are usually polymerized.

Table: Typical cyclic monomers and their usual polymerization mechanism.

Name	Structure	Ring size	Mechanism
Olefin	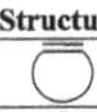	4,5,8	Metathesis
Ether	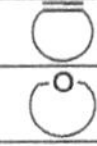	3–5,7	Cationic, anionic
Thioether	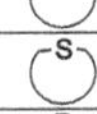	3,4	Cationic, anionic
Amine	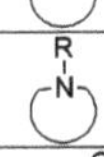	3,4,7	Cationic
Lactone	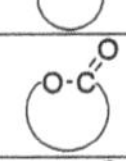	4,6–8	Anionic, cationic
Thiolactone	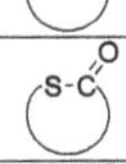	4–8	Anionic, cationic
Lactam	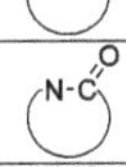	≥4	Anionic, cationic
Disulfide	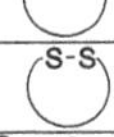	≥8	Radical
Anhydride	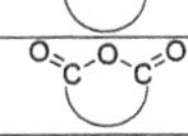	5 and ≥7	Anionic
Carbonate	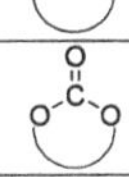	6–8 and ≥20	Anionic
Silicone	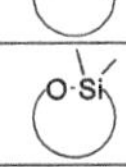	6,8 and ≥10	Anionic, cationic
Phosphazene	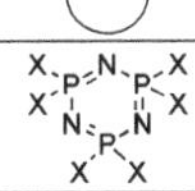	6	Cationic
Phosphonite	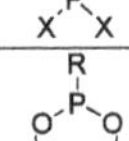	3,5–7	Anionic

Radical Ring-opening Polymerization

Via a ring-opening polymerization, especially radical ROP (RROP), it is possible to produce polymers with the same or lower density than the monomers. This is interesting for applications where it is desirable to maintain a constant volume during the polymerization such as tooth fillings, coatings and accurate molding of electrical and electronic components. RROP offers novel routes to polyesters and polyketones particularly using vinyl substituted cyclic monomers. A less common group of monomers which undergo RROP are the bicyclobutanes, whereby the bridge across the four-membered ring can be viewed as a concealed double bond.

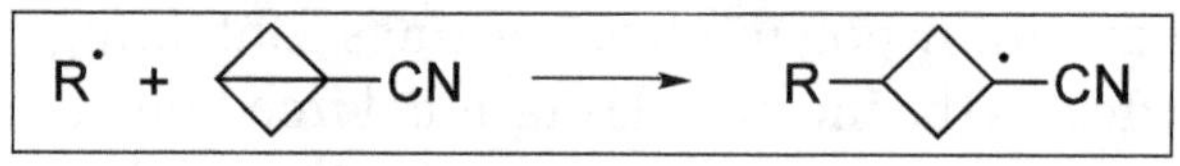

The radical ring opening polymerization (RROP) of a bicyclobutane.

Vinyl Substituted Cyclic Monomers

Vinyl cyclopropane and its derivatives have been studied by several groups. The three membered ring has considerable ring strain (115 kJ mol−1) and the simple RROP thus competes with the vinyl chain polymerization.

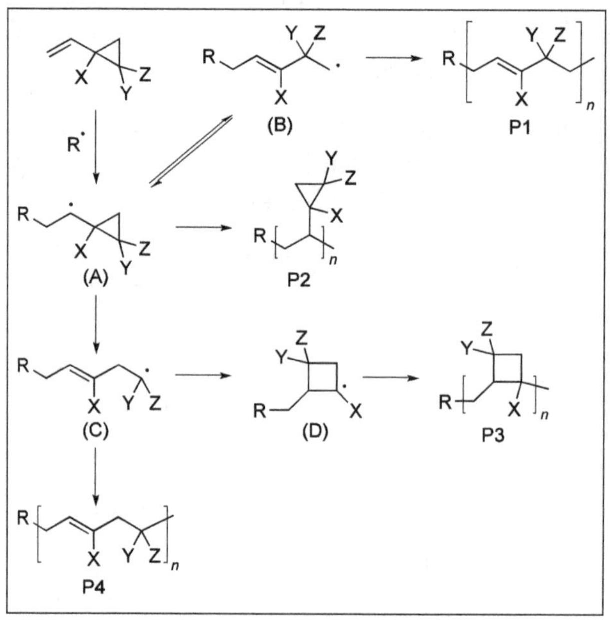

The RROP of vinyl cyclopropane showing the competing reaction pathways.

Which of the simple ROP products P1 and P4 predominates depends on the substituents Y and Z and the preference of the ring to open at the CX–CH2 or the CX–CY bond. The simple vinyl chain polymerization P2 and the intermediate formation of a four-membered ring to yield P3 are additional possibilities. Vinyl oxiranes yield polyethers via ROP.

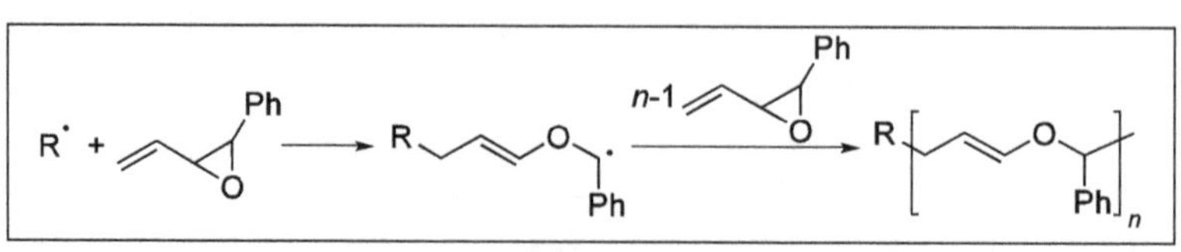

The RROP of phenyl (Ph) vinyl oxirane.

Vinyl cyclobutanes can also undergo RROP.

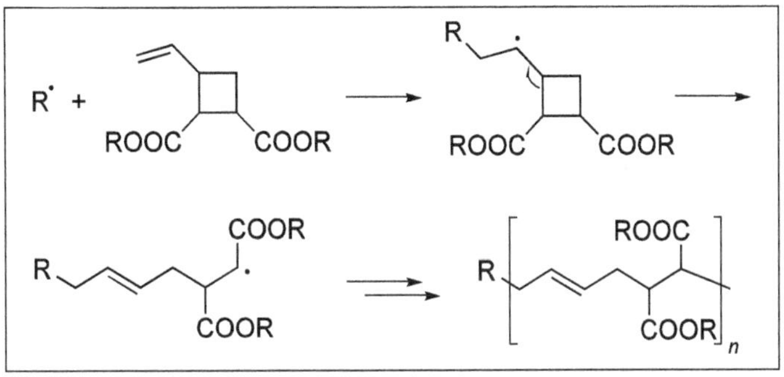

The RROP of a substituted vinyl cyclobutane.

Methylene Substituted Cyclic Monomers

The ring-opening of ketene acetals has proved to be a novel route to useful polyesters although the sensitivity of the monomers to acid hydrolysis makes their handling difficult.

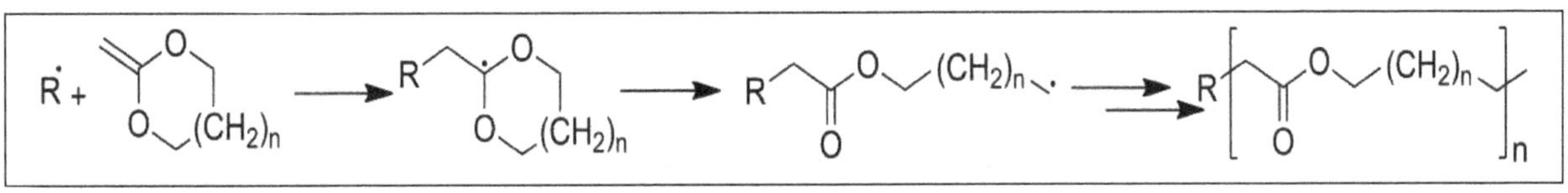

The RROP of ketene acetals.

The polymerization of 2,2-diphenyl-4-methylene-1,3-dioxolane yields a polyketone rather than a polyester since benzophenone is eliminated by each ring-opening step.

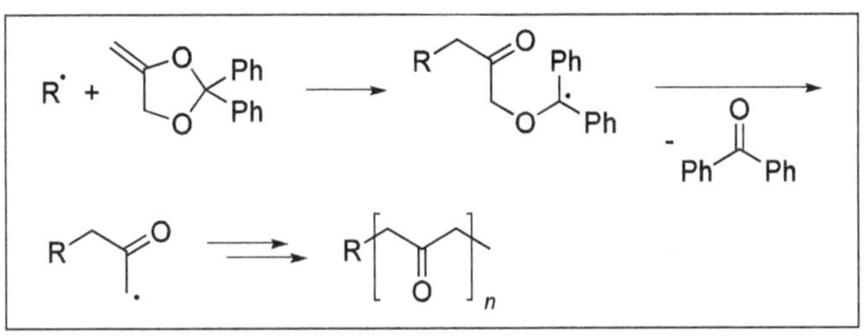

The radical polymerization of 2, 2-diphenyl-4-methylene-1,3-dioxolane to yield a polyketone.

Heterocyclic monomers with nitrogen or sulfur heteroatoms have also been shown to undergo radical ROP.

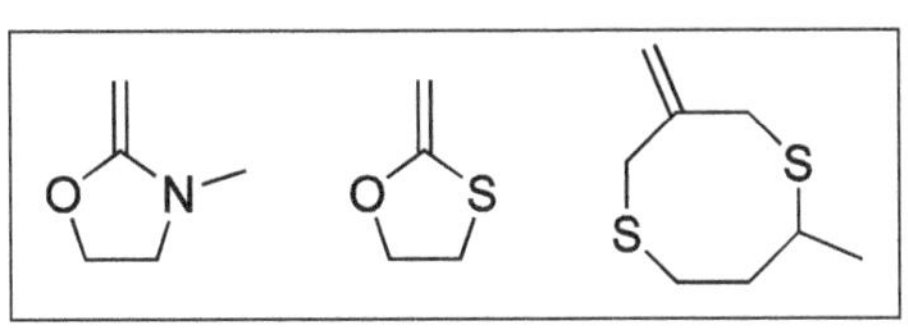

Some heterocyclic monomers which have been shown to undergo radical ROP.

Double Ring-opening

Spiro ortho-carbonates and their analogues have been the subject of considerable interest due to the volume expansion which occurs when the crystalline monomers are polymerized (If the monomers are first melted then a volume contraction takes place). A typical reaction scheme is shown in figure.

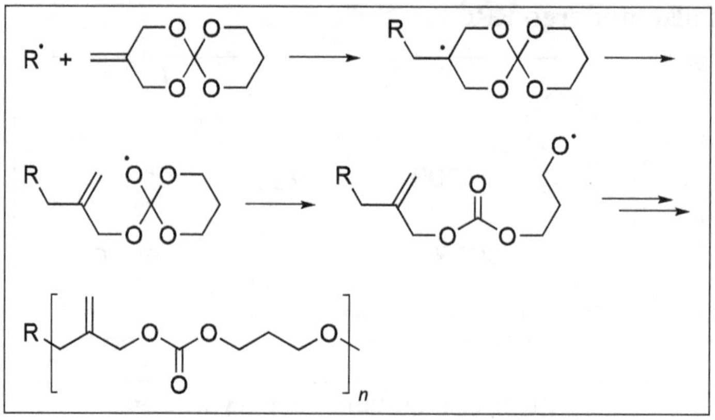

The radical ROP of a spiro carbonate.

The spiro ortho esters shown in figure have also been shown to polymerize, essentially quantitatively, via a radical ROP.

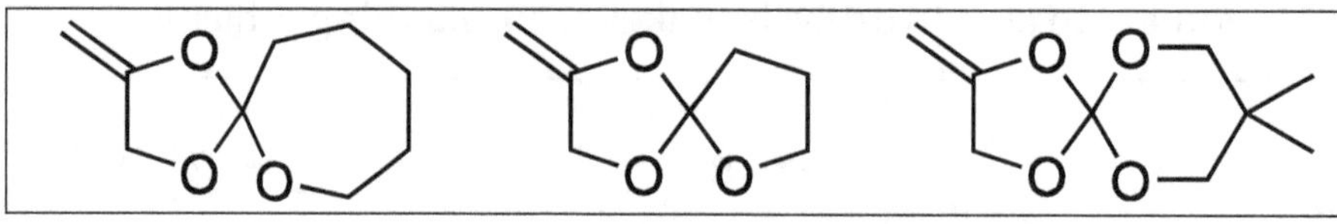

Some other examples for the radical ROP spiro monomers.

Degradable Polyesters via Radical Ring-opening Homo- and Copolymerization

The polyester synthesis under 2.2 is not a typical CROP, AROP or RROP in that a rearrangement takes place as each ring opens. Despite the apparently promising work of Bailey the synthesis of polyesters via a RROP of ketene acetals initially received little interest. However, with the increasing interest in biodegradable polyesters and the versatility of RROP with respect to the structures accessible via very different monomers and their copolymerization with conventional vinyl monomers has changed the situation significantly.

Figure shows an example of such a copolymerization. It is clear that by varying the monomers or their proportions, or indeed by varying the mode of initiation, e.g., NMP (nitroxide mediated polymerization], RAFT (reversible addition fragmentation transfer polymerization) or ATRP (atom transfer radical polymerization) many and varied polymers can be obtained, optimized for their use in agricultural, medicinal or pharmaceutical fields.

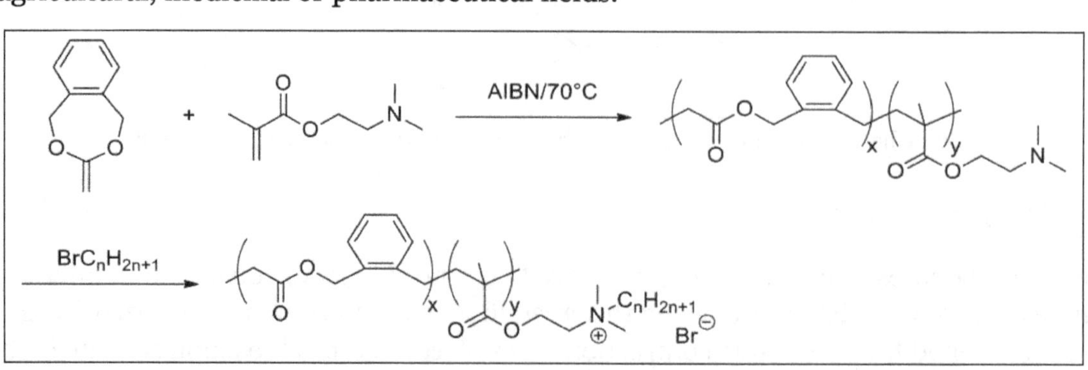

An example of a radical ring opening copolymerization of a ketene acetal.

Cationic Ring-opening Polymerization

Ring-opening polymerizations involving a positively charged intermediate (Cationic ROP or CROP) provide several important industrial polymers. These include polyacetals, copolymers of 1,3,5-trioxane and oxirane or 1,3,5-trioxane and 1,3-dioxolane, polytetrahydrofurans, copolymers of tetrahydrofuran and oxirane, poly (3,3-bis(chloro-methyl)oxetanes), polysiloxanes, polymers of ethyleneimine and polyphosphazene.

There are essentially two mechanisms for CROP: The first involves a growing chain with a cationic center at the chain end which adds to a monomer molecule via a S_N1 or S_N2 mechanism, whereby X represents an oxygen, a nitrogen or a sulfur atom.

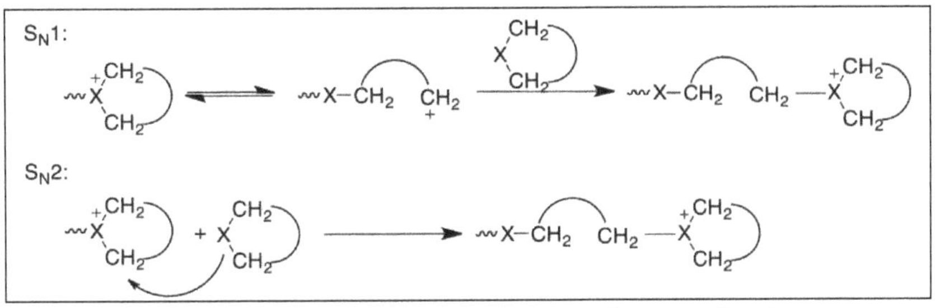

CROP involving active chain ends; S_N1 and S_N2 mechanisms.

The polymer structure can be further controlled by the use of appropriate side groups. Thus, the exchange of an ether side group for an ester or carbamate results in a different, well defined, microstructure during the CROP of cyclic dithiocarbonates with methyl triflate initiator due to effective suppression of the isomerization of the monomer and the stabilization of the growing chain via the ester (carbamate) moiety.

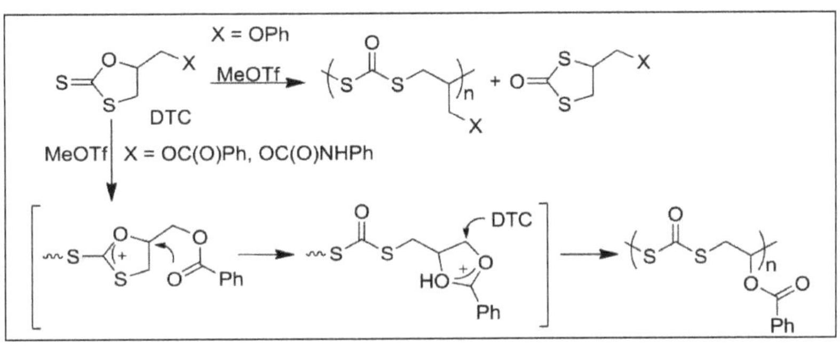

Controlling the reaction path of a CROP of dithiocarbonates using Neighboring Group Participation (NGP).

An extreme example of NPG is reported by Suzuki *et al.* TfOH who describe the polymerization of an aromatic lactone in which the lactone is part of a fused ring system (3-Isochromanone). Using a variety of spectroscopic techniques and an elegant analysis, the authors suggest that the resulting polymer, which could be prepared in excellent yield via a bulk polymerization (2 h at 120 °C, 5 mol % triflic acid (TfOH) as catalyst), is formed via an intermediate Friedel-Crafts alkylation step as a result of the stabilization of the growing chain end in the form of a benzyl cation.

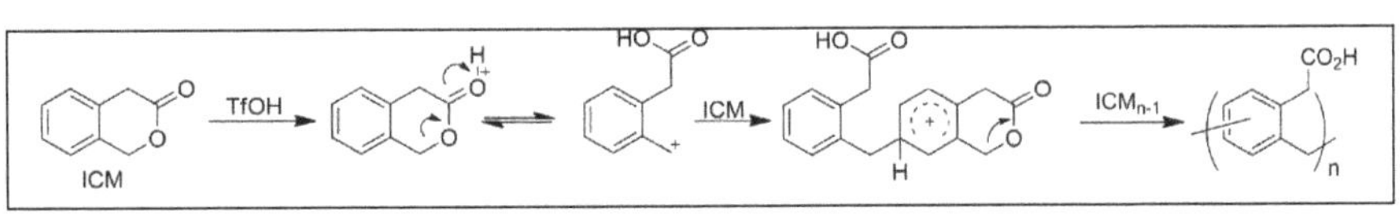

The CROP of 3-isochromanone.

A treatment of the propagation of CROP which can be considered as resolved (vide infra), specifically concerning, e.g., 1, 3-dioxolanes, as to whether the propagating species are linear tertiary or secondary ions and thus, the mechanism is one of ring expansion or of linear growth. In terms of the microstructure of the polymers produced this is not trivial. In a very detailed analysis Plesch *et al.* showed that under conditions where the concentration of water was less than 10^{-4} mol L^{-1} and at relatively low monomer concentration, the secondary oxonium ion predominates and the ring expansion mechanism prevails. In contrast to the suggestions of Plesch, Jaacks and later Penczek postulated, based on extensive experimental results that propagation takes place essentially via open chains. Finally, from even more detailed studies it was found that ring-expansion is favored at low monomer concentrations and the open chain mechanism dominates the propagation at higher monomer concentration.

An alternative, third, mechanism proposed for the propagation of CROP is that the monomer is activated and carries the cationic center; the growth reaction is an electrophilic attack of the activated monomer on the chain end.

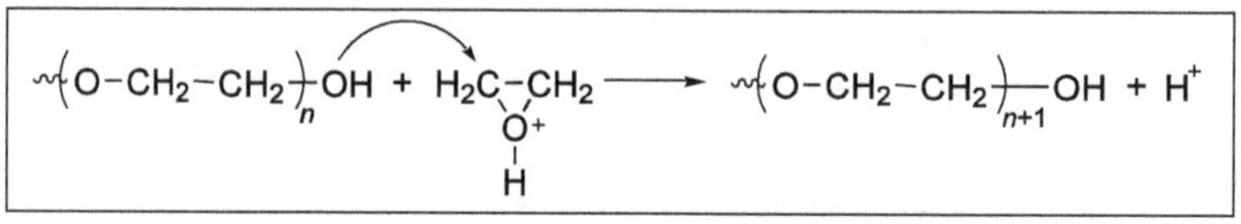

CROP via an activated monomer mechanism.

For some systems it has been shown that both mechanisms operate simultaneously.

There are at least three types of reactions, which compete with the propagation reaction during a CROP as shown in figure.

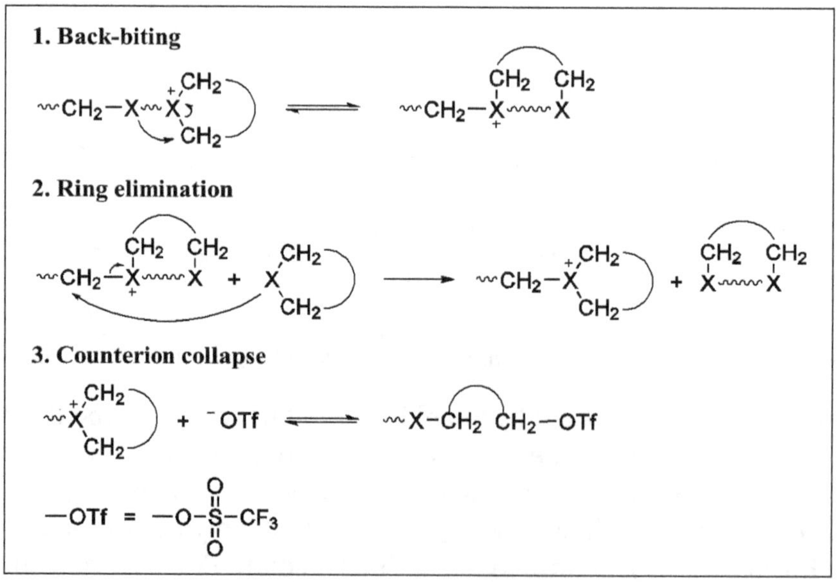

Three principal side reactions during a CROP.

Initiation

CROP can be initiated by Brönsted acids, carbenium ions, onium ions, photoiniators and covalent initiators.

Brönsted Acids

Protons can initiate the polymerization of numerous heterocyclic monomers, e.g., oxiranes, 1, 3-dioxolanes, thiiranes, aziridines, tetrahydrofurans, lactones, 1, 2 oxazolines and cyclic siloxanes as shown in figure assuming an activated chain end mechanism, in analogy to figure. "Dry" acids such as HCl, H_2SO_4, as well as $HClO_4$ and $HOSO_2CF_3$ have frequently been used.

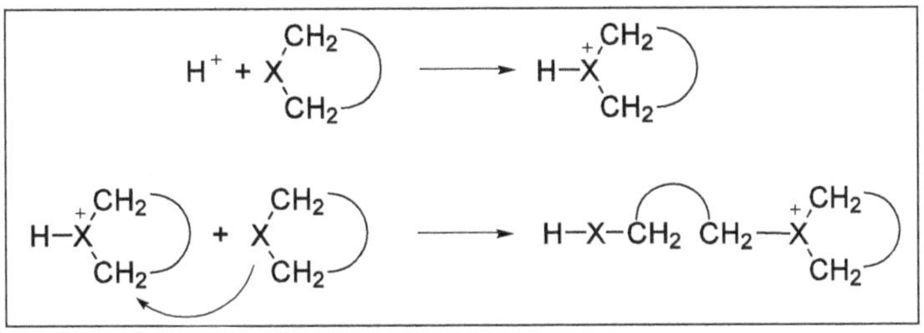

Initiation of a cationic ROP by a Brönsted acid.

Alternatively, the heteroatom poly-acids of the general structure shown in figure can be employed as initiators.

$H_nX_mMt_zO_y$ X = P, Si
Mt = Mo, W, V

General structure of heteroatom poly-acids.

The acids with X = P can be relatively easily prepared from phosphoric acid and the required metal oxide; they are stable in air and are soluble in a variety of organic solvents. Two examples are well documented in the: $H_3PMo_{12}O_{40}$ and $H_3PW_{12}O_{40}$.

The CROP of epoxides and oxetanes via Brönsted acids are often characterized by an induction period and, although not completely explained, the more recent studies of Crivello's group modifying the kinetics of the photo initiated CROP by the addition of crown ethers (which are similar in structure to the oligomers formed in the early stages of polymerization) or 2,6 di-tert-butyl pyridine do offer a plausible explanation for some of the observed effects. Thus, control of an auto accelerated thermal polymerization is improved by such additives.

Carbenium Ions

Initiation of CROP by various carbenium ions has been frequently studied, the general reaction scheme is shown in figure ($R^+ = (C_6H_5)_3C^+$, $(C_6H_5)_2CH^+$, $(C_6H_5)CH_2^+$, $H_2C=CH–CH_2^+$ and $(CH_3)_3C^+$). The carbenium ions can react with rings of various structures and having various hetero atoms:

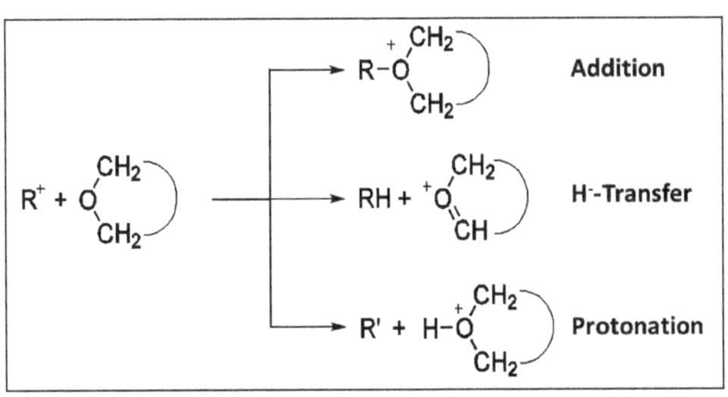

Initiation of CROP by carbenium ions (R = $R^+(-H^+)$).

Onium-ions

Onium-Ions can be prepared in a separate reaction or *in situ*. Initiation by onium ions follows, in a general sense, the scheme shown in figure; typically the reaction involves the transfer of a carbenium ion and can thus be seen as an extension of the initiation by carbenium ions (vide supra).

Typical examples are the reactions of trialkyl oxonium ions with cyclic acetals, ethers, or sulfides as well as with amines.

A variation on this type of initiation is reported by Li and Yu who have reported the use of glycosyl ortho-alkynyl benzoates and a gold triflate salt (PPh_3AuNTf_2) to prepare poly-THF with glycosyl terminal groups.

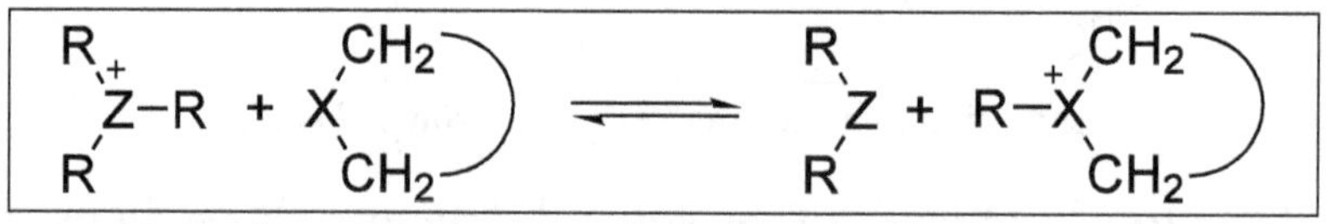

Initiation by onium-ions (Z and X may be identical, e.g., oxygen).

Covalent Initiators

The polymerization of 2-Alkyl-2-oxazolines can be initiated by classical alkylating reagents, such as benzyl bromide.

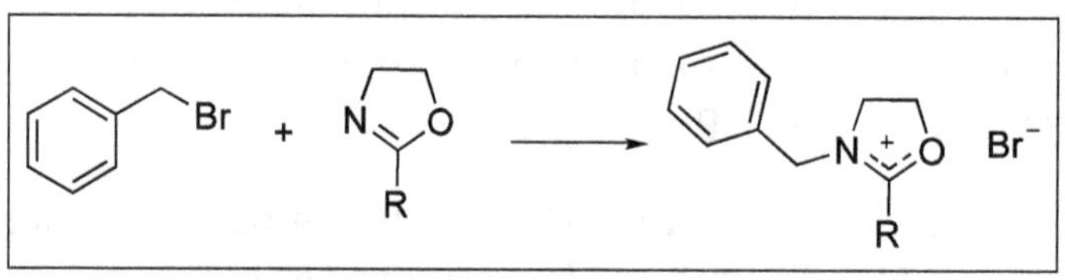

Initiation by covalent reagents.

Esters of Brönsted acids (e.g., ROSO2CF3 and ROSO2F) can also be used as initiators, whereby the covalent initiators act as precursors for a carbenium ion initiation (vide supra). In particular, the use of triflic acid anhydride is worth noting as a route to growing polymers of THF with two active ends.

Lewis Acids

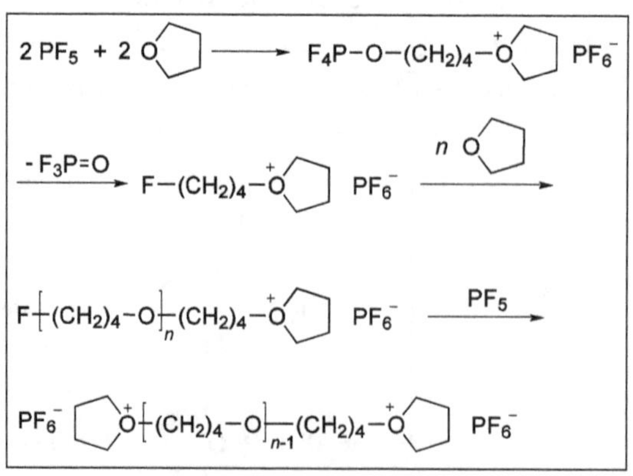

Initiation of the CROP of THF by PF_5. Chain growth from both chain ends.

These compounds require a coinitiator (activator), e.g., an alkyl halide or water, in order to polymerize heterocyclic monomers. The most frequently employed Lewis acid for the cationic polymerization is BF_3 and its precursor BF_3OEt_2. However, in order to apply them for the ROP of THF the addition of cationogenic compounds such as acid halides is required. Detailed studies have also been made using PF_5, e.g., for the polymerization of THF for which a mechanism supported by ^{19}F and ^{31}P NMR spectroscopy has been proposed.

Photoinitiators

Sulfonium- and iodonium-salts have achieved industrial importance as cationic photo initiators. Both heterolytic and homolytic reaction mechanisms are possible. Taking sulfonium salts as an example, via a heterolytic process the sulfonium salt is raised to a singlet-state, which then relaxes to give Ph_2S and Ph^+ (from Ph_3S^+). The Ph^+ moiety can then react with e.g., an alkyl group (this may be the solvent, polymer or monomer) to yield a proton and PhR.

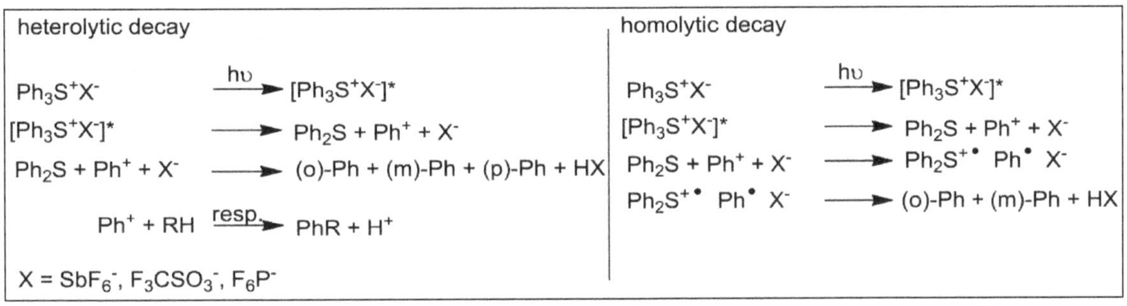

Formation of an initiator by heterolytic or homolytic splitting of a photoinitiator (*: Singlet-state).

Via the heterolytic splitting of the initiator can be considered as an acid generator or photoacid generator (PAG). Typical examples are triphenylsulfonium triflate and diphenyliodonium triflate. The latter has been studied recently to selectively cleave the carbon-oxygen bonds in a vinyl functionalized oxasilaspirocycle to form both homo- and copolymer polysiloxanes via a double ring-opening CROP.

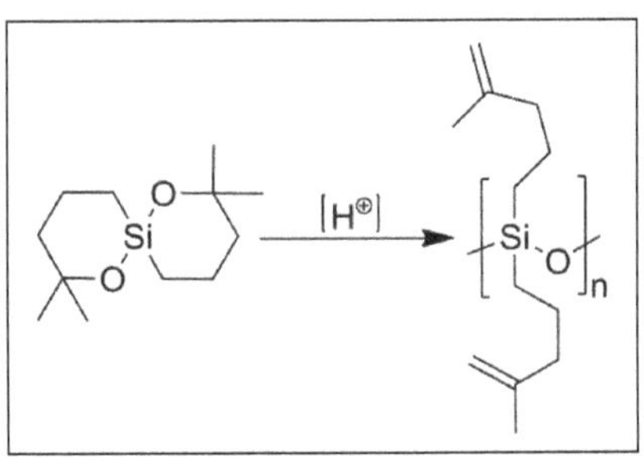

The formation of a polysiloxane via a double ring-opening cationic polymerization.

The homolytic splitting of triphenylsulphonium cations leads to a phenyl radical, which then undergoes a Fries photo-rearrangement together with the diphenylsulphonium cation radical to yield by-products such as *p*- und *o*-(Phenylthio)biphenylenes. That some m-substituted products are observed can be explained by a heterolytic decomposition followed by an electrolytic aromatic substitution of the diphenylsulphonium cation radical. Photo initiators have proved useful as crosslinking agents for hole (defect) conductors (e.g., aromatic amine polymers which can give up an electron to yield a radical cation, an electronic defect) with oxetane side groups, which are employed in the manufacture of OLED. The reaction schemes are shown in figures.

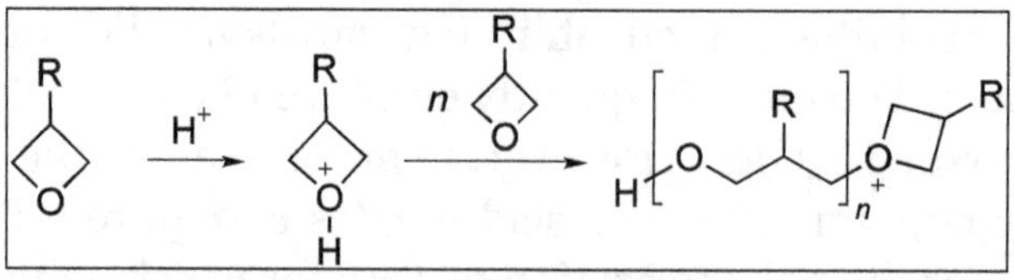

CROP of oxetanes (R = Polymer chain with oxetane side groups).

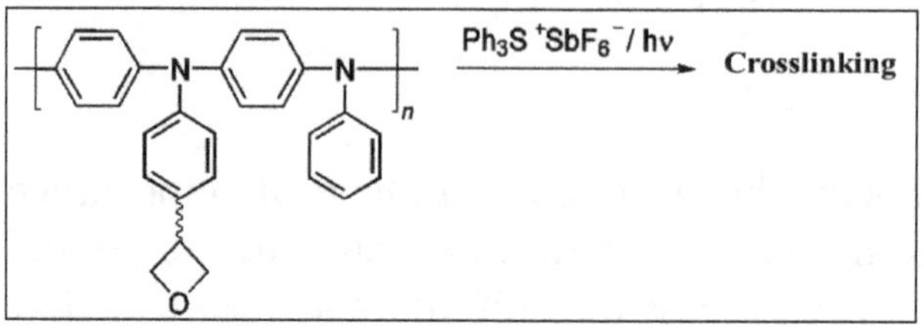

The cationic crosslinking of a hole defect polyarylamine.

If the hole defect polymer has numerous oxetane side groups, the ROP leads to crosslinking and the polymer becomes insoluble. The soluble polymer can be coated (e.g., by spin-coating), dried and then photochemically crosslinked so that further coating is possible without dissolving the deposited film.

The solubility of the photo initiator can be improved by increasing the length of the side chain.

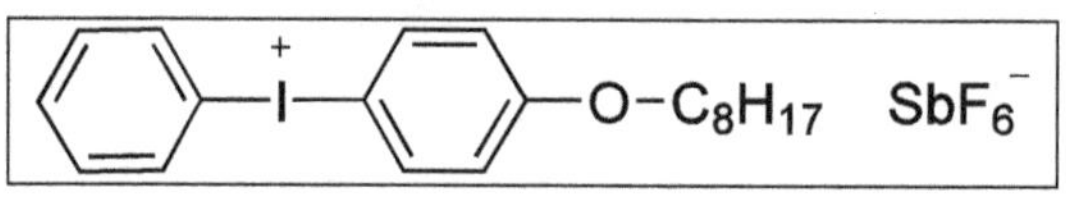

Structure of an iodonium photo initiator.

Enzymes have also been shown to initiate the ring-opening polymerization of oxazolines (EROP). By employing different sugar molecules having a methyl oxazoline side group a variety of natural and synthetic polysaccharides could be produced, in particular some of the most complicated synthetic polymeric structures and synthetic chitin. The fundamental mechanism, although the enzyme interacts with the intermediates and increases the reaction rate considerably, remains a CROP. It is worth noting that polyoxazolines have been shown to be biocompatible and potentially useful as carriers of pharmaceutically active agents. An additional group of initiators for the CROP of oxazolines are the activated metallocene complexes described by Kourti *et al.* In this study, zirconium and hafnium metallocenes were activated with pentafloroborates or their anilinium salts to initiate a CROP of 2-methyl and 2-phenyl oxazolines. That the polymerization is indeed a CROP was corroborated by appropriate blank experiments. However, since the metallocene does not take part in the propagation step the steric control associated with metallocene polymerizations cannot be expected for these systems and the polymerizations appear to show no special kinetic or mechanistic features. However, the authors do note an increase in the MWD and M_n at close to complete conversion, which may be explained by backbiting reactions.

Chain Growth

The active species in CROP are oxonium, sulfonium, ammonium or phosphonium ions.

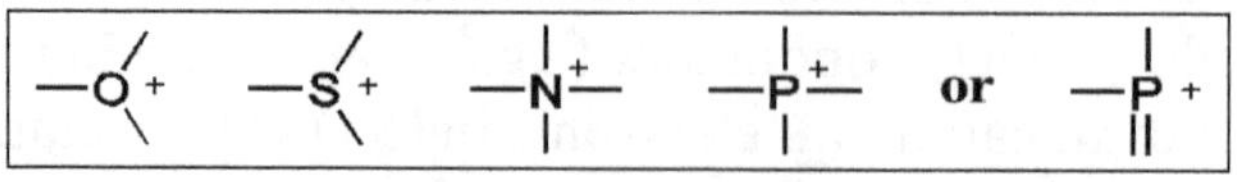

The structure of the growing chain ends in CROP.

The active species can be identified by NMR and can be quantified e.g., by reacting with e.g., sodium phenolate to yield a UV active end group.

The structure of the growing chain ends during CROP for a variety of monomers:

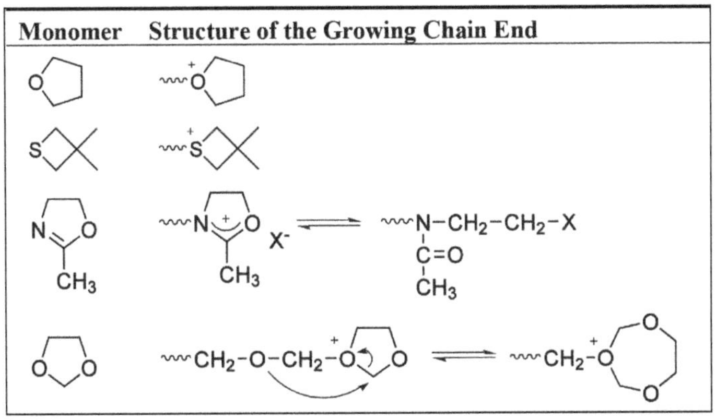

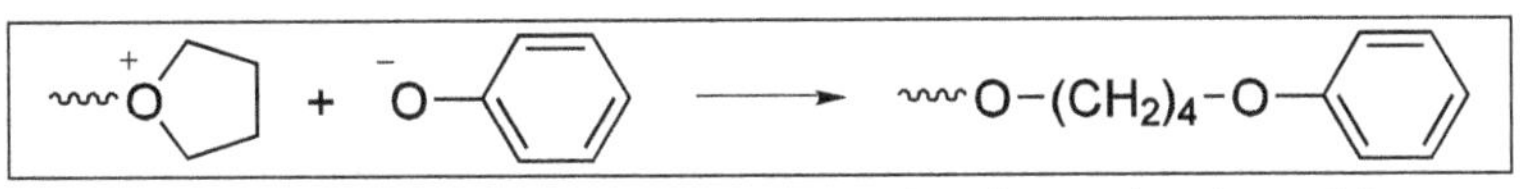

Termination of CROP by phenolate to allow quantitative determination of the growing chains.

An alternative method for the quantitative determination of the active chain ends is to react these with, e.g., triphenyl phosphine and to analyze the solution by ^{31}P NMR. It is worth noting that these polymers are not suitable for further analysis, as they tend to degrade although the solutions are stable, for days at least, provided they are kept under an inert atmosphere.

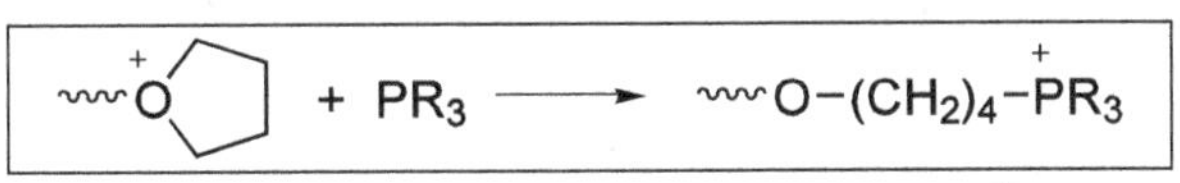

Reaction of the CROP growing chain ends with a phosphine to allow quantitative analysis.

Propagation via an activated monomer can be visualized as shown in figure.

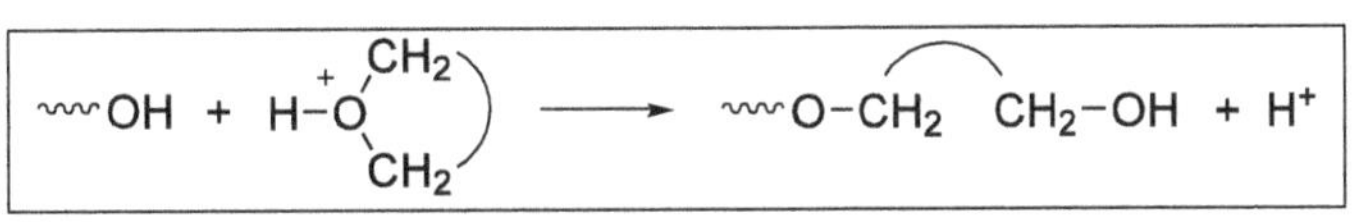

Chain growth via an activated monomer.

With asymmetric monomers chain growth via an activated monomer rather than an activated chain end can yield different polymer microstructures as shown in figure.

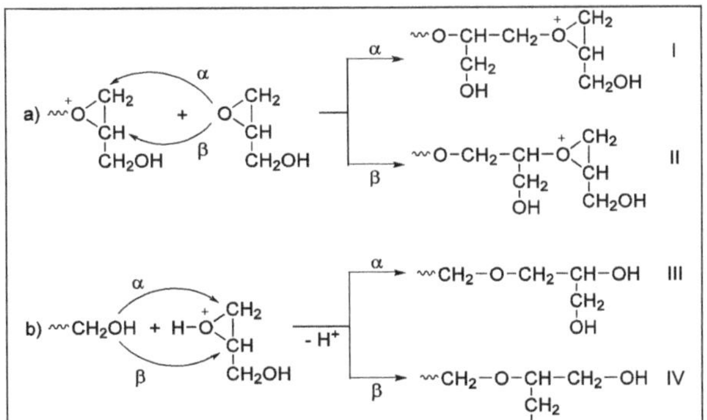

The microstructure of poly-glycidol: (a) Growth via an active chain end and (b) via an activated monomer mechanism.

The experimental results suggest that the predominant structure is III so that it can be assumed that for this monomer the activated monomer mechanism predominates.

The CROP of hexamethylene cyclotrisiloxane (D3) is an important industrial process generally catalyzed by trifluoromethane sulphonic acid or metal salts of this acid. For the CROP of cyclic siloxanes two propagation reactions have been proposed.

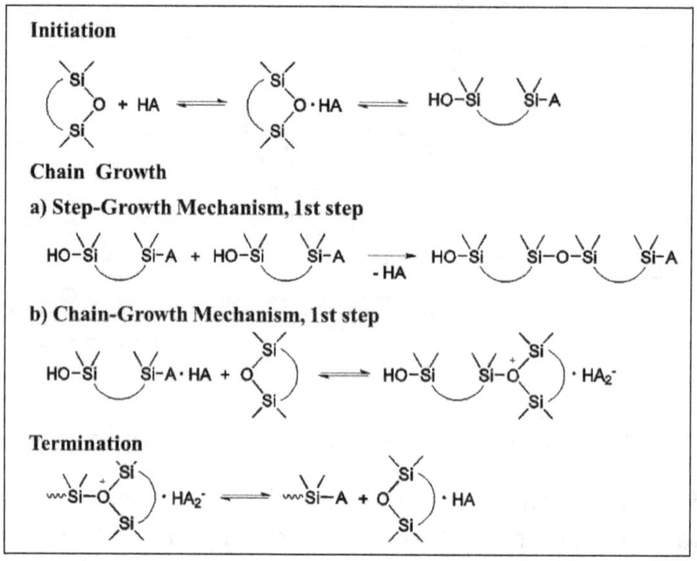

Two alternative propagation routes for the ROP of cyclic siloxanes: (a) Step-growth and (b) Chain-growth.

In contrast to the chain growth polymerization of olefins, where the termination leads to an active species being destroyed, the termination of cyclic monomers containing a heteroatom (such as the siloxane shown in figure) leads in the first instance to the chain growth being stopped but the resulting monomer unit can start a new chain as long as monomer is available. Thus, the distinction between the terms transfer and termination becomes gradual. Since the reaction intermediates are extremely reactive.

Termination

The CROP of some heterocycles, especially cyclic ethers such as THF and oxetane can, at appropriate temperatures and monomer concentrations, be considered as "living" (i.e., without termination). Intentionally adding termination reagents may e.g., be done to enable the end groups to be (quantitatively or qualitatively) analyzed. Examples are the addition of phenoxy anions or phosphines but polyanions can also be used to yield block copolymers. These reactions are summarized in figure.

$Na^+ \, {}^-[St]_n^- \, Na^+$	+ $2[THF]_m^+PF_6^-$	$\longrightarrow$	$[THF]_m$-$[St]_n$-$[THF]_m$
$2[St]_n^- \, Na^+$	+ $PF_6^- \, {}^+[THF]_m^+ \, PF_6^-$	$\longrightarrow$	$[St]_n$-$[THF]_m$-$[St]_n$

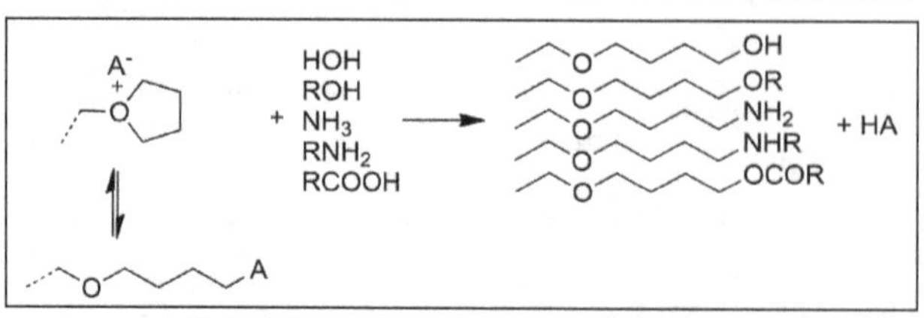

Intentional termination of CROP.

However, backbiting and/or intermolecular alkylation is intrinsic to such systems.

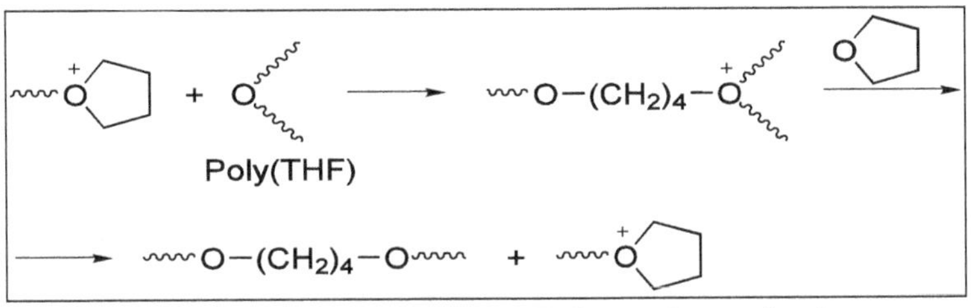

Transfer (Termination) via an alkyl transfer mechanism.

Since these reactions usually only become predominant when the monomer becomes depleted, only then do they represent chain termination rather than transfer; the charged species cease to grow due to lack of monomer and the second stage of the scheme in figure does not occur.

In some systems, backbiting seems to be negligible and the systems appear to be living. In this work the authors report the CROP of 3-benzyl-1,3-oxazolidine-2-thione (BOT) to yield a polythiourethane using BF_3OEt_2 in nitrobenzene solvent and although initiator efficiencies are low (*ca.* 0.1 based on the amount of BF_3OEt_2 employed), the M_w/M_n measured was *ca.* 1.0 for all initiator/monomer ratios studied and both di- and triblock copolymers could be produced with 3-benzyl tetrahydro-1,3-oxazolidine-2-thione (BTOT) as a second monomer. This work is especially worth citing for its use of an unusual end-capping agent ($Et_2NCS^-Et_2NH_2^+$), which can then be analyzed by NMR, the use of a variety of analytical techniques to confirm the structures as well as for the excellent yields of polymers with improved thermal stability and perhaps interesting optical properties via a CROP. The kinetic analysis also corroborates the idea (*vide supra*) that the ROP is driven by the ring strain; thus, the rate of propagation for the 5 membered BOT ring is twice that of the 6 membered ring in BTOT, although the authors ascribe the difference to the different steric hindrance around the thiocarbonyl groups of the two monomers.

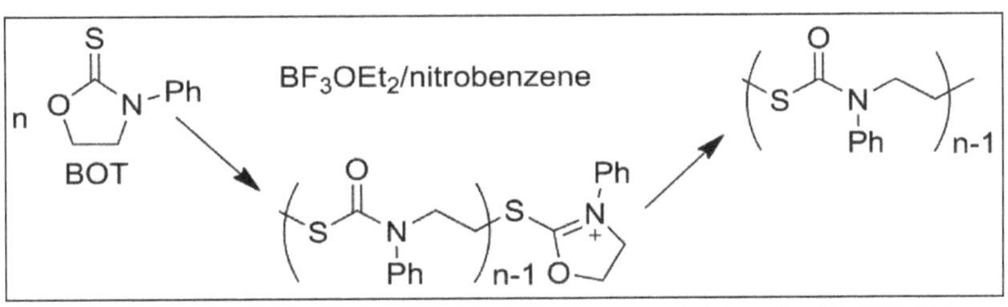

The CROP of 3-benzyl-1,3-oxazolidine-2-thione initiated with BF_3OEt_2.

Figure shows this reaction for the ROP of oxirane. The resulting dioxonium ring, in this case, is inert to further chain growth.

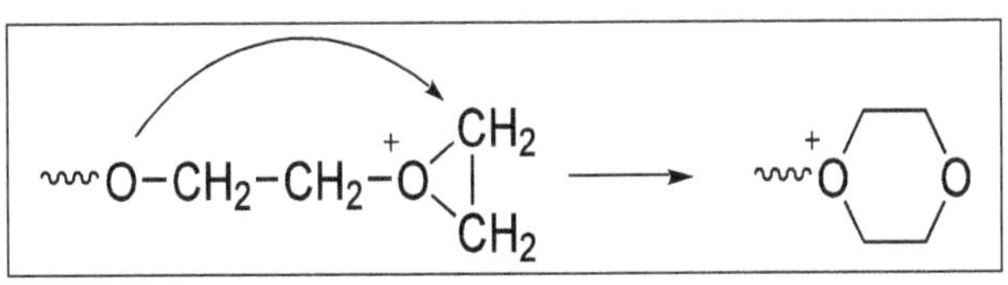

Termination via back-biting; ring expansion.

The collapse or decomposition of the counter ion also leads to a termination of chain growth, whereby this reaction may be an equilibrium reaction and thus, the chain ends considered as dormant rather than terminated.

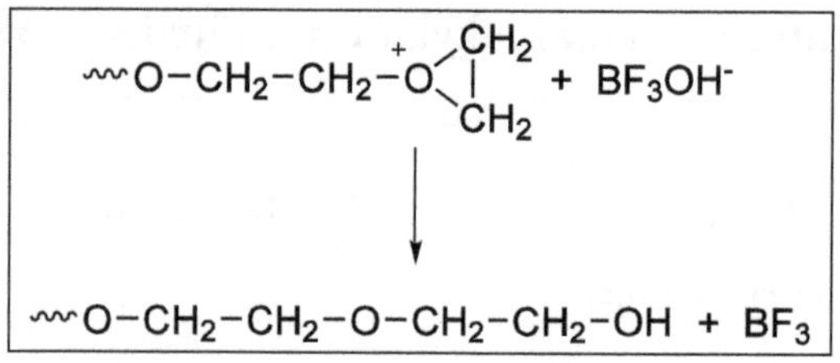

Termination of chain growth due to collapse of the counter ion.

The living nature of CROP has been utilized to form block copolymers from 2-methyl-2-oxazoline, 2-ethyl-2-oxazoline, 2-n-nonyl-2-oxazoline and, e.g., styrene or acrylates. The CROP of the oxazoline was terminated by a carboxylate functionalized chain transfer agent and the second block formed by a reversible addition-fragmentation chain transfer (RAFT) polymerization.

CROP reactions are often rather slow and reaction times even at higher temperatures are often of the order of hours. The reaction rate can be increased considerably (up to 400 fold) by the use of microwave rather than simple heating, although it has been shown that the effect is a simple thermal effect.

Anionic Ring-opening Polymerization

Anionic ROP (AROP) can be described as the nucleophilic attack of the growing chain end on a heterocyclic monomer molecule. The AROP of ε-caprolactam to produce Nylon 6 is an important industrial process.

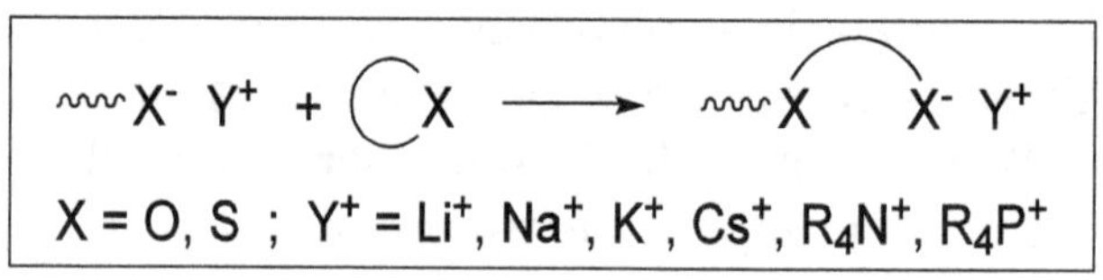

The AROP as a nucleophilic addition reaction.

Some of the heterocycles that can be polymerized anionically are listed in table with their corresponding active chain ends.

Table: Anionically polymerizable heterocycles and the corresponding active species.

Monomer	Structure of the Growing Chain
	~~ $CH_2-CH_2-O^-$
	~~ $CH_2-CH_2-S^-$
	~~ $S-CH_2-CH_2-CH_2^-$
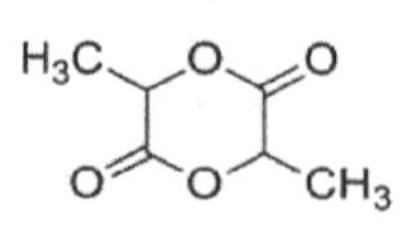	~~ $O-C(=O)-CH(CH_3)-O-C(=O)-CH(CH_3)-O^-$
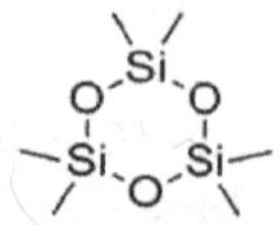	~~ $Si-O-Si-O-Si-O^-$

Initiation

Some of the more important initiators for AROP are listed in table.

Table: Initiators for an AROP of heterocycles.

Description	Structure	Monomer
Radical anion	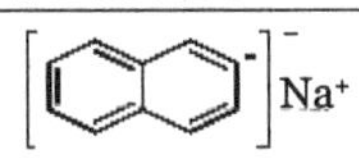	Ethylene oxide
		Propylene sulfide
Carbanion	C_2H_5–Li^+	Thietane
	n-C_4H_9–Li^+	Propylene sulfide Hexamethylcyclo-trisiloxane
	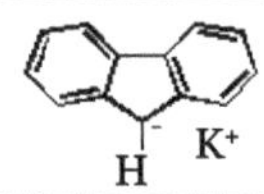	Ethylene oxide
Alcoholate	$CH_3O^-K^+$	Styrene oxide
		β-Propiolactone
Silanolate	$(CH_3)_3SiO^-K^+$	β-Propiolactone
		ε-Caprolactone
Carboxylate	$CH_3COO^-K^+$	β-Propiolactone
		Ethylene oxide
		Propylene oxide
Thiolate	$C_2H_5S^-K^+$	Propylene sulfide
Lactam anion	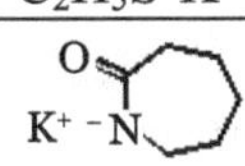	ε-Caprolactam
Amine	$(C_2H_5)_3N$	Propylene sulfide
Al-trialkoxide	Al(O-*t*-C_3H_7)$_3$	ε-Caprolactone
Al-Dialkyl monoalkoxide	$(C_2H_5)_2AlOCH_3$	Lactide ε-Caprolactone

Propagation

For unsymmetrically substituted rings there are formally two ways the monomer ring can be opened as shown in figure.

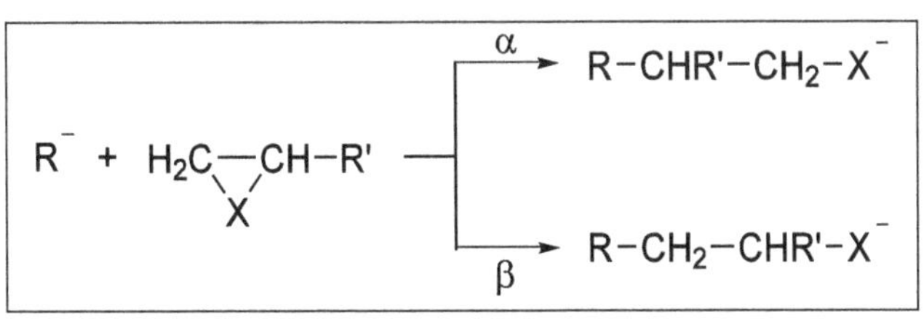

Nucleophilic ring-opening of asymmetrical heterocycles.

Although not trivial, with careful microstructure and end group analysis the preferred reaction route can be identified. Additionally, simple consideration of the stability of the reaction intermediates can prove helpful. Thus, the carboxylate ion (route β) is preferred, even for lactones, and the nucleophilic attack takes place at the least substituted carbon atom.

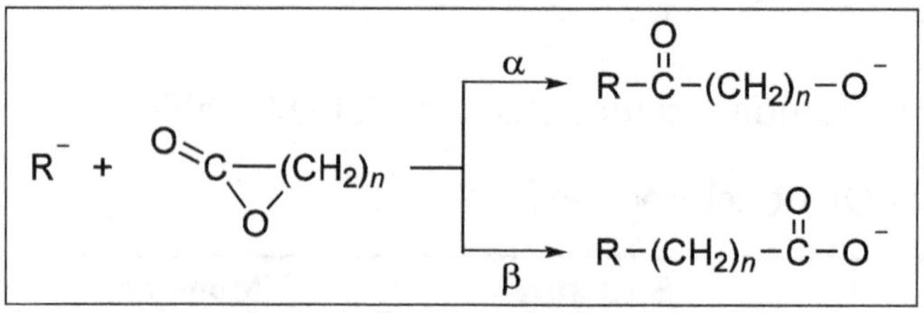

The AROP of lactones.

An additional possibility for determining whether route α or β is preferred is to short stop the polymerization e.g., with diphenyl chlorophosphate and then to analyze the solution with ^{31}P NMR.

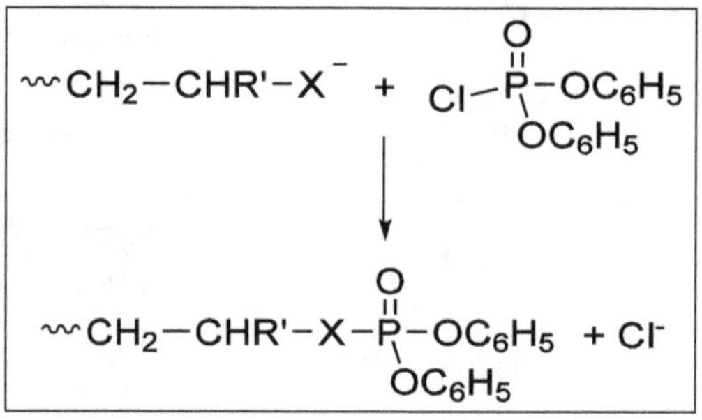

Polymerization short stop with diphenyl chlorophosphate (Alcoholate (X = O) or thiolate (X = S)).

The AROP of lactams can be initiated with strong bases such as the alkali metals, metal hydrides or metal amides which give, as a first step, lactam anions.

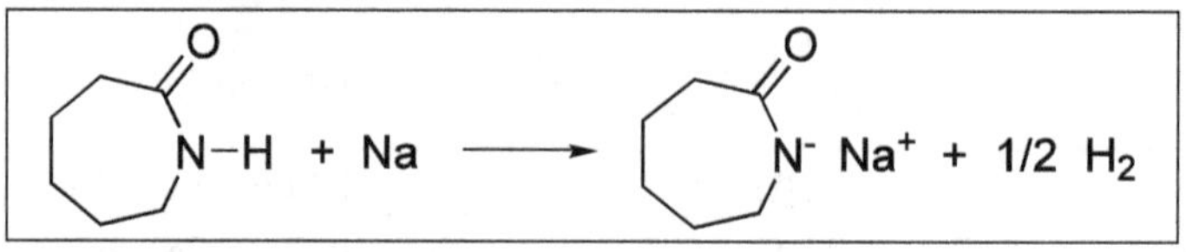

The formation of a lactam anion from the reaction of sodium with ε-caprolactam.

The lactam ring is then opened in the second step.

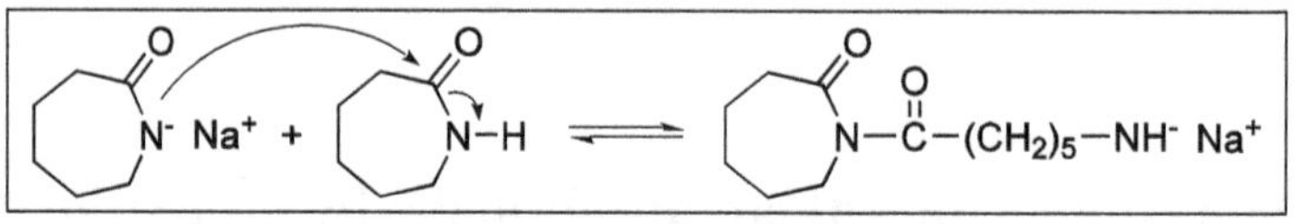

The opening of a lactam ring by a lactam anion.

The next step is the proton abstraction by the very reactive primary amine anion from another monomer to regenerate a lactam anion (Activated monomer mechanism).

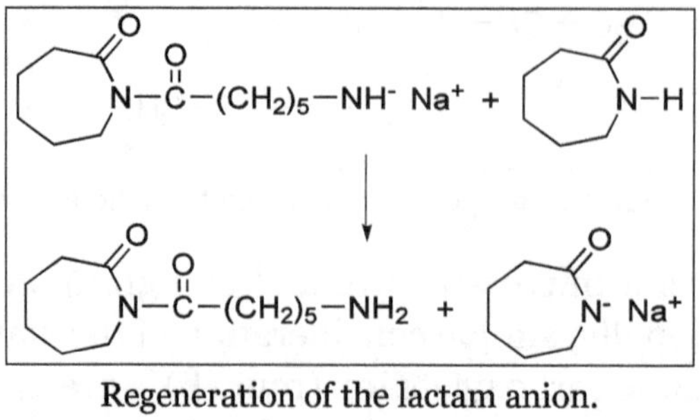

Regeneration of the lactam anion.

Whereby the terminated dimer can also react with a lactam anion and thus grow.

An AROP of methyl oxirane (1,2-propylene oxide) is an industrial process for the preparation of oligomers in combination with oxiranes (ethylene oxide). The extensive chain transfer to monomer, typical of substituted oxiranes prohibits the formation of high polymers; the reaction is shown in figure. It should be noted that the allyl alkoxide ion can also generate propagating chains. This leads to less than ideal alcohol group formation where simple end capping reactions are employed (or diol formation with bifunctional initiators).

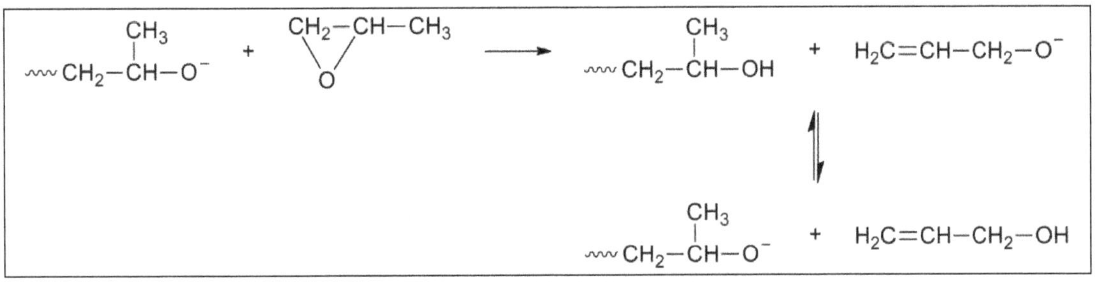

AROP of methyl oxiranes with chain transfer to monomer.

A particularly interesting case of AROP is the condensation ROP of a Leuchs' anhydride (*N*-Carboxy-α-amino acid anhydride), which eliminates a molecule of CO_2 for each addition step.

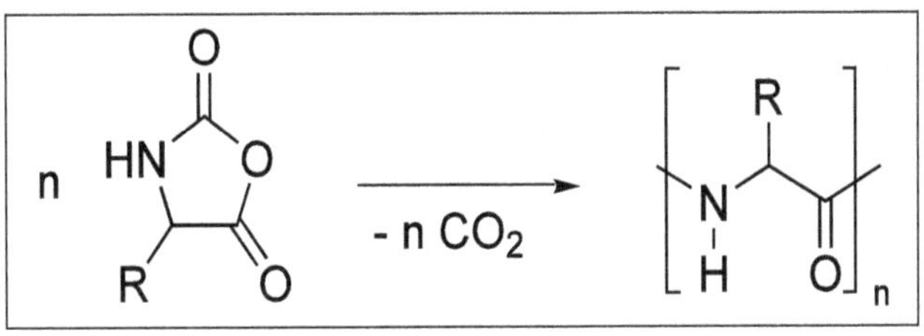

Formation of "Nylon-2" from a Leuchs' anhydride.

As initiator, e.g., primary amines can be employed. It could be shown by using ^{14}C labeled monomer that the CO_2 contains exclusively the C-2 carbon atom. Thus, the free electron pair of the primary amine nitrogen attacks the monomer at C-5, the carbon with the lowest electron density and the CO_2 elimination results from sequentially breaking the C-5–O-1 and the C-2–N-3 bonds, whereby the amine nitrogen acts as the base.

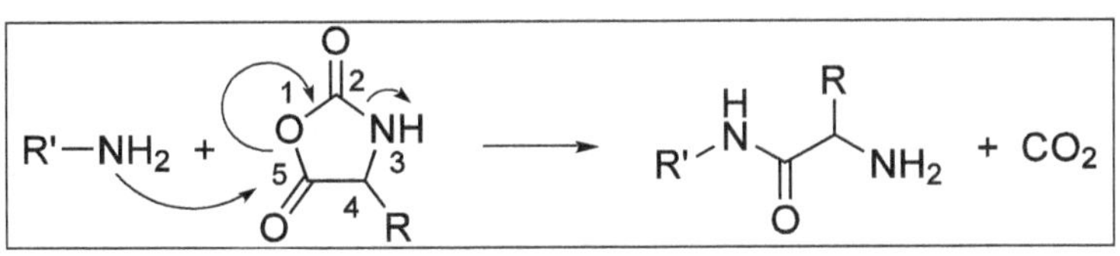

Mechanism of CO_2 elimination as a result of the attack of a primary amine on a Leuchs anhydride.

An alternative to a covalent initiator is the use of a base B^- (e.g., $^-OCH_3$, H^-, $^-OOCCH_3$) which generates an anion from oxazolidin-2, 5-diones as shown in figure. This anion can then initiate chain growth via a nucleophilic attack on another monomer.

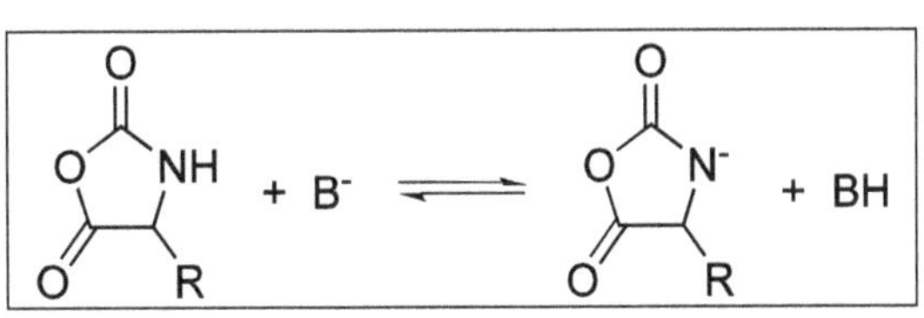

Formation of an anion from a Leuchs' anhydride using a base.

Cyclic siloxanes can also undergo ROP via anionic intermediates employing alkali metal oxides or hydroxides as initiators. On an industrial scale, octamethylenecyclotetrasiloxane (D_4) is employed as a starting material. For the propagation step the $\Delta H_p \approx 0$ kJ mol^{-1}, but due to the increased rotational freedom of the side chains the $\Delta S_p = +6.7$ J mol^{-1} polymerization does occur. However, depolymerization competes with polymerization and rings and chains are in equilibrium. The equilibrium is shifted to the ring form with increasing size of the substituents. If other siloxanes with different (e.g., phenyl or methyl) substituents are mixed with D_4 statistical copolymers are obtained. An alternative, employing D_3 (Hexamethylenecyclotrisiloxane) as the monomer, has the disadvantage that the synthesis is more elaborate.

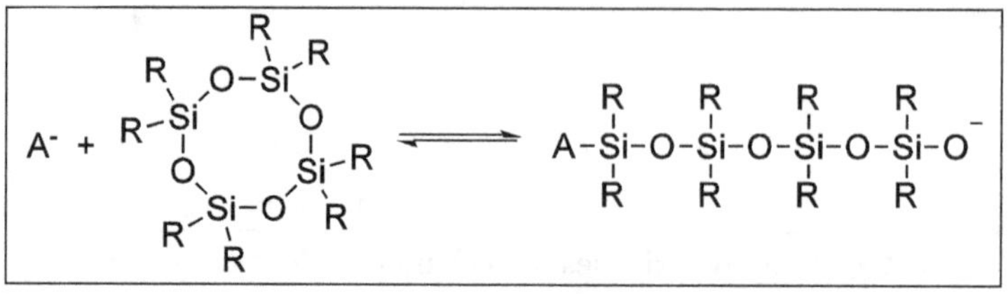

Equilibrium between rings and chains during the AROP of cyclic siloxanes ($A^- = \sim Si(R)_2O^-$).

Transfer and Termination

Chain termination reactions are more accurately described as transfer reactions as long as additional monomer is available. The active centers of AROP, e.g., alcoholates or carboxylates are not only nucleophilic but act as bases and can abstract protons from the monomer to initiate new chains. Thus, the polymerization of propylene oxide initiated with alkali metal salts does not give polymers with high molar masses (Vide supra). One possibility to increase the molar mass of the products is to add crown ethers as complexing agents for the counter ions to the polymerization system. The free-ion/ion-pair equilibrium is shifted to the free-ion side; the free-ions preferentially add to monomer rather than abstract a proton.

Ring-Opening Metathesis Polymerization (ROMP)

Olefin metathesis is a reaction that can be seen to have influenced both organic and polymer chemistry more than any other during the last few decades. Historically, the ring-opening metathesis reaction has its origin in polymer chemistry: In 1955 Anderson and Merckling (DuPont), working with norbornene and $TiCl_4$/EtMgBr catalysts obtained a polymer containing a highly unsaturated backbone which some years later was identified as the product of a ring-opening polymerization.

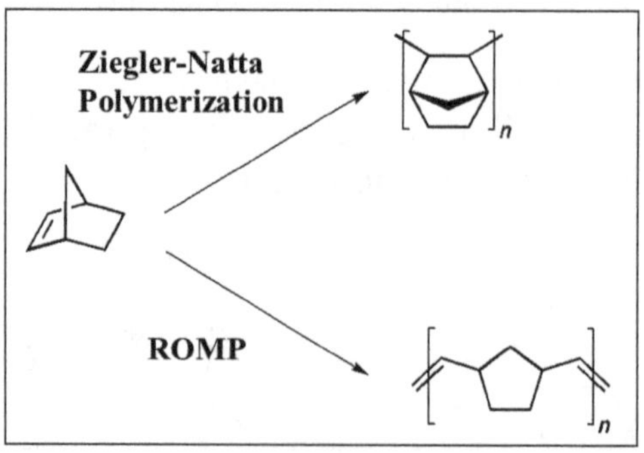

Ziegler-Natta polymerization and ring-opening metathesis polymerization (ROMP) of norbornene.

The development of ROMP is also documented in the academic. The driving force for ROMP is the release of the ring strain enthalpy and, since the number of molecules is considerably reduced by polymerization, a positive ($-T\Delta S$) must be compensated by the enthalpy term in order for the reaction to proceed according to:

$$\Delta G = \Delta H - T\Delta S$$

Table lists ΔH-, ΔS- and ΔG-values for a selection of cyclic olefins. The positive ΔG^o confirms that cyclohexene cannot be polymerized by a ROMP at room temperature.

The remaining double bonds in the backbone of the polymer allow backbiting reactions to form macrocycles, which often observed as products from the ROMP of cyclic olefins.

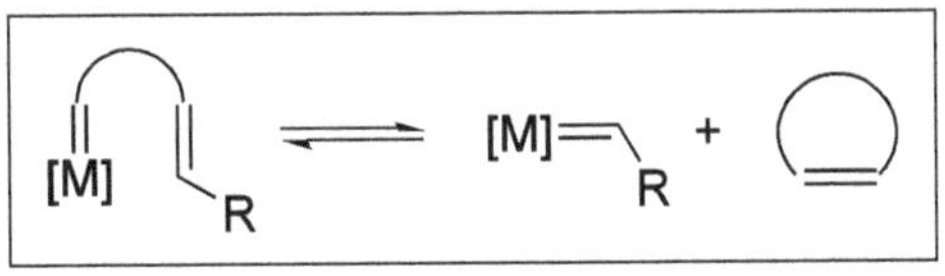

The formation of macrocycles via back biting during the ROMP of cyclic olefins.

Table: Thermodynamic parameters for the ROMP of a selection of liquid monomers to amorphous polymers at 25 °C.

Monomer	Configuration of the double bond in the polymer	ΔH°/(kJ/mol)	ΔS°/(J/(mol K))	ΔG°/(kJ/mol)
Cyclopentene	Cis trans	−15.4 −18	−52 −52	−0.3 −2.3
Cyclohexene	Cis trans	−1−3 1−3	−31 −28	6.2 7.3
Cycloheptene	Cis trans	−16 −20	−20 −17	−10.0 −15.0
Cyclooctene	Cis trans	−20 −22	−2 −2	−19.4 −21.5
1, 5-Cyclooctadiene	Cis trans	−25 −33	−5 −5	−23.5 −31.5

The metathesis of olefins and in particular cyclic olefins can be initiated by homogeneous, heterogeneous and immobilized catalysts. The classical catalysts are the oxides of Group VI transition metals on Lewis acid carriers, which are robust and can be regenerated. Typical examples of heterogeneous catalysts are WO_3/Al_2O_3 and MoO_3/Al_2O_3, as homogeneous catalysts $WCl_6/SnBu_4$ und $MoCl_6/SnPh_4$ are quoted whereby these and a large number of other systems. Only with the development of well-defined complexes based on Molybdenum or Ruthenium metals could a tolerance with respect to e.g., acid, ester and amide functionalities be developed and the possibility for a much wider variety of polymer structures opened up. Schrock, Grubbs and Chauvin were honored with a Nobel Prize in 2005 for their synthetic and mechanistic work in this area. A selection of catalysts is shown in figure.

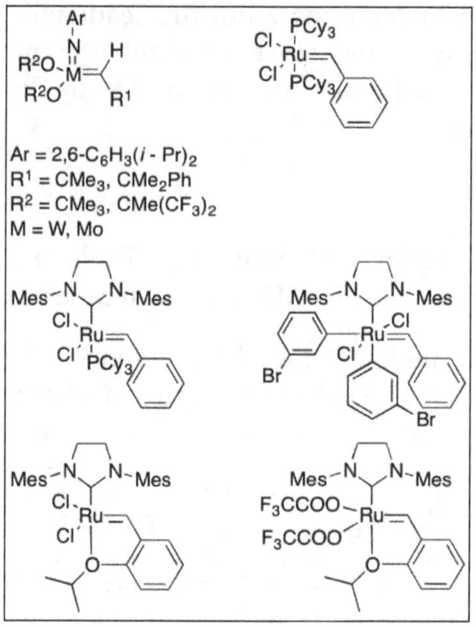

A selection of ROMP catalysts.

ROMP is employed to produce a variety of interesting polymer structures of which a selection is shown in figure together with their reported commercial applications.

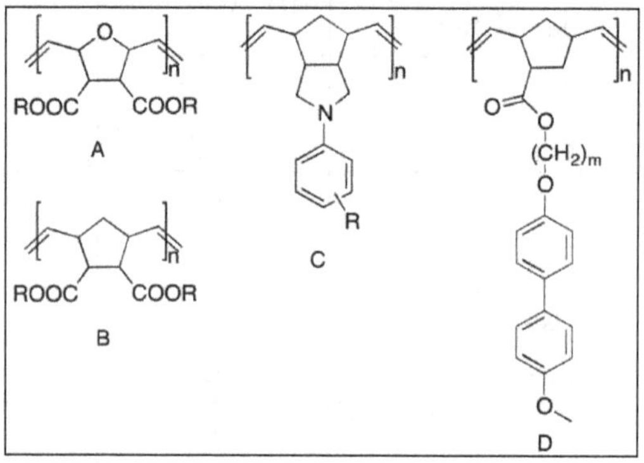

Functional polymers via ROMP: (A) Super absorber, after crosslinking; (B) Ion exchange resin, flocculation aid; (C) High temperature thermoplast; (D) Polymer with liquid crystal side chains (m ≥ 20).

The best known industrial products synthesized via ROMP are: Norsorex® (Polynorbornene, Astrotech Advanced Elastomerproducts GmbH), Vestenamer® (Polycyclooctene) (Evonik Degussa GmbH), polycyclopentene and polycyclopentadiene (Metton®) are all well established commercial products.

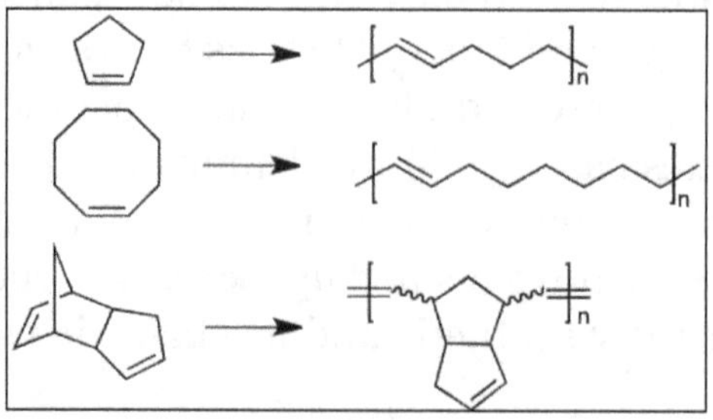

The structures of Norsorex®, Vestenamer® and Polycyclopentadiene.

It is now widely accepted that the critical propagation step in ROMP is the formation of a metallocyclobutane intermediate from a coordination of a cycloolefin by the catalytic metal center of the carbine complex followed by a cycloaddition. The metallocyclobutane then undergoes a cycloreversion and the metal atom separates from the olefin bond.

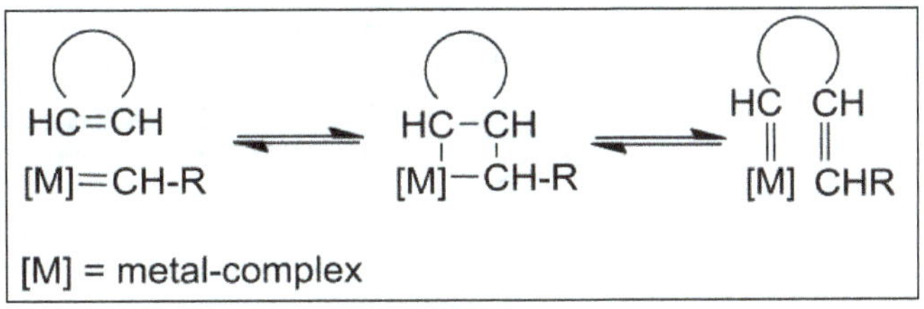

The mechanism of ROMP including the metallocyclobutane intermediate.

The reactivity of the metal complexes with respect to different functional groups is graphically shown in figure; thus, ruthenium complexes react preferentially with olefins and can, indeed, be used in aqueous media for metathesis reactions whereas the titanium complexes are sensitive to almost all impurities.

Activity ↑ / Tolerance ↓	Titanium	Tungsten	Molybdenum	Ruthenium
Activity	Alcohols, water	Alcohols, water	Alcohols, water	**Olefins**
	Acids	Acids	Acids	Alcohols, water
	Aldehydes	Aldehydes	Aldehydes	Acids
	Ketones	Ketones	**Olefins**	Aldehydes
	Ester, amides	**Olefins**	Ketones	Ketones
Tolerance	**Olefins**	Ester, amides	Ester, amides	Ester, amides

The reactivity/tolerance of metathesis catalysts.

One disadvantage of homogeneous metathesis catalysts is that they cannot be reused; one solution is to immobilize them on either inorganic or organic carrier materials as shown in figure.

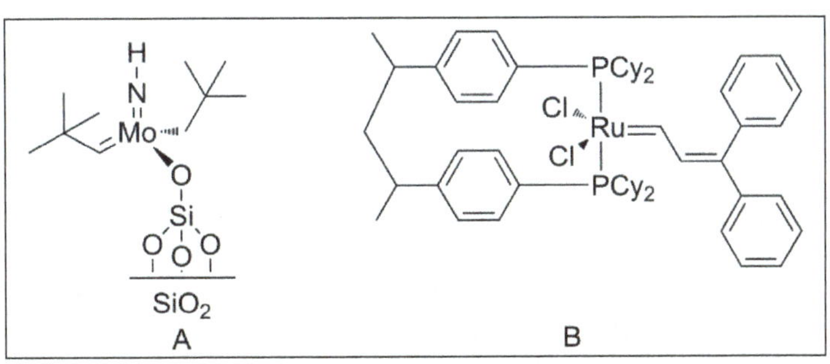

Metathesis catalysts immobilized on A silica (Schrock Type) and B polystyrene (Grubbs type).

The opposite of ROMP, i.e., a ring "closure" metathesis polymerization occurs during the metathesis of of α, ω-dialkynes. Thus, controlled backbiting reactions occur leading to interesting new polymers with perfect 5- or 6-membered ring structures which exhibit good electrical conductivity, as shown in figure.

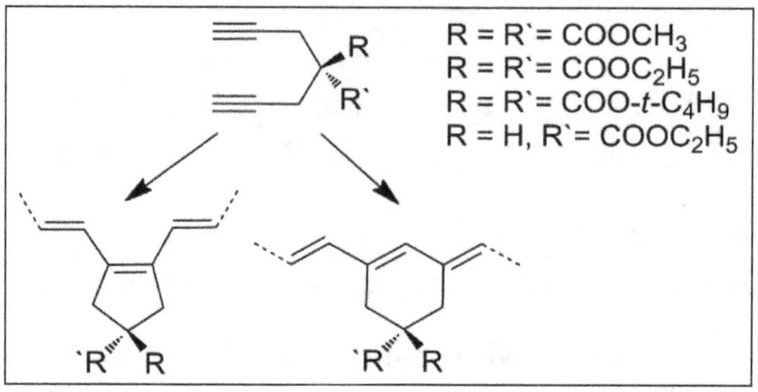

The ring-closure metathesis polymerization of α, ω-dialkynes.

The good control of polymer structures possible via ROMP has led to many studies concerning the synthesis of polymers with well-defined end groups.

Other Ring-opening Polymerizations

The pioneering work of Inoue involving the ROP of oxiranes initiated by aluminium porphyrin/alcohol describes an "immortal" polymerization; a living polymerization during which the growing chains do not "die". Rapid and reversible chain transfer takes place between the growing chains and alcohol molecules. Thus, only a very small initiator concentration is required. The number of polymer chains is equal to the sum of the number of initiator and transfer molecules. The overall reaction is shown in figure using the ROP of trimethylene carbonate, initiated by a [Zn]-mediated initiator in the presence of an alcohol (R^2OH) as an example.

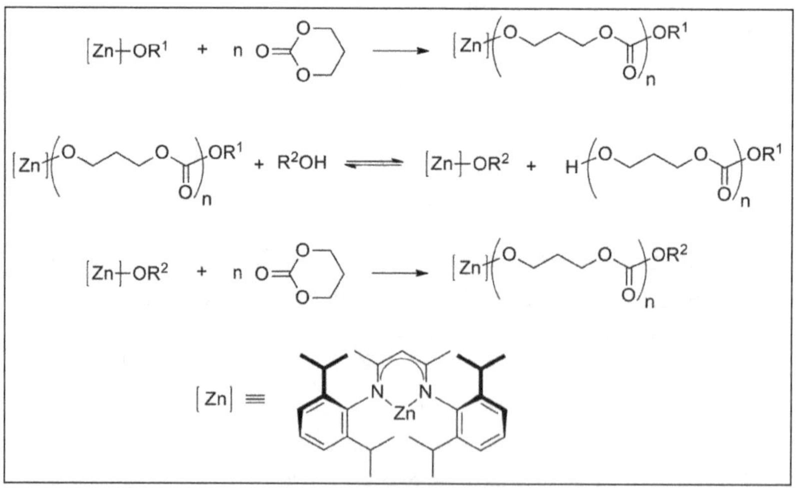

[Zn] mediated living ring-opening polymerization of trimethylene carbonate.

The initiators based on porphyrin complexes or complexes with related structures based on Al, Zn, Y, Lu, In have been used for the polymerization of e.g., oxiranes, thiiranes, lactones, lactides, cyclic carbonates and others. Early examples of alternating copolymerization are shown in figure.

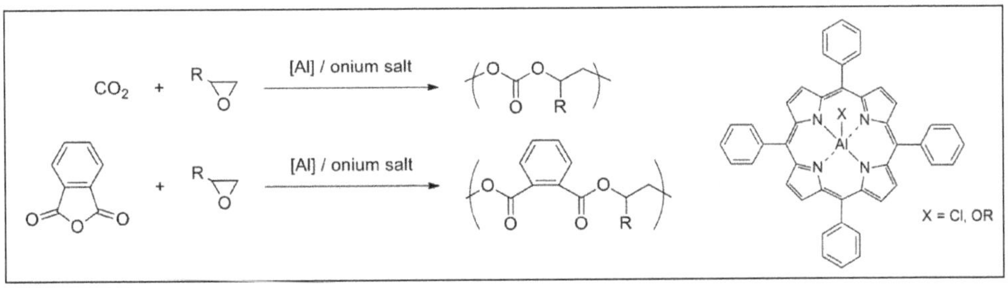

Alternating copolymers via a living copolymerization of oxirane with CO_2 or phthalic anhydride.

Furthermore, a great variety of block copolymers have been synthesized using this method.

One typical example is presented in figure.

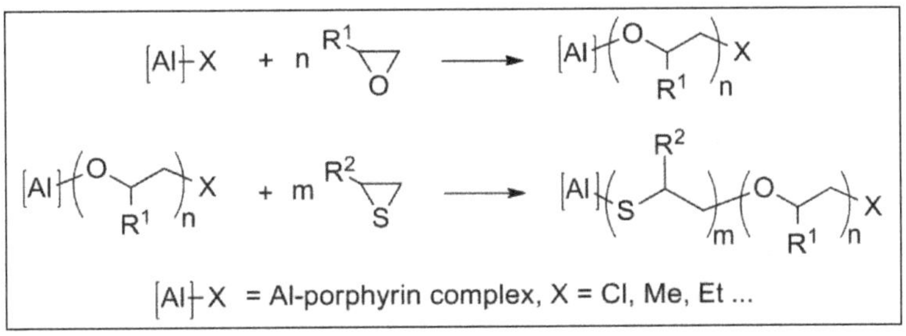

Synthesis of a block copolymer via a living ring-opening polymerization.

Cyclic phosphonites undergo an electrophilic ring-opening polymerization via an Arbuzov-type reaction to yield a polyphosphonate.

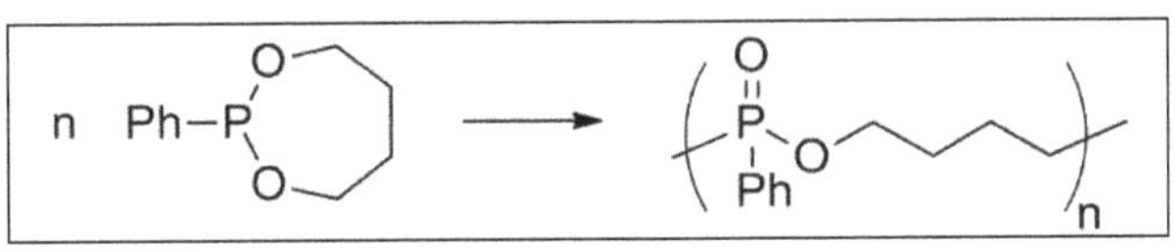

Electrophilic ring-opening polymerization of a cyclic phosphonite.

During the polymerization the phosphorus atom of the monomer is oxidized (from +3 to +5) and the carbon of the CH_2O group is reduced (from −1 to −3) to yield the PCH_2 moiety in the polymer; *i.e.*, an intramolecular oxidation-reduction reaction takes place.

An interesting aspect of this type of ROP is the alternating copolymerization of a cyclic phosphorus (III) compound as reductant with a simple vinyl monomer as oxidant, as depicted in figure.

This copolymerization takes place without any initiator or catalyst.

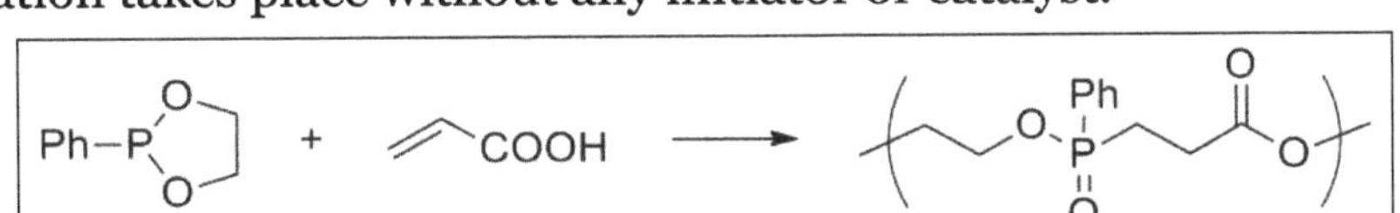

Intermolecular reduction-oxidation during the alternating copolymerization of a cyclic phosphorus (III) monomer with acrylic acid.

Ring-opening Polymerization of Phosphazenes

An interesting and promising class of polymers, which is still waiting for a real industrial break-through, can be obtained by the ring-opening polymerization of phosphazenes. The contains reports on more than 300 polymers on this basis. The simple polymerization of $(PNCl_2)_3$ is carried out in sealed vessels under an inert atmosphere by heating to *ca.* 250 °C for several hours. A conversion of much greater than 50% is avoided to minimize the formation of insoluble material. The polymerized material is then converted, by reaction with a stoichiometric amount of an alcoholate, an amine or an organometallic reagent, to yield a polymer more stable to hydrolytic decomposition.

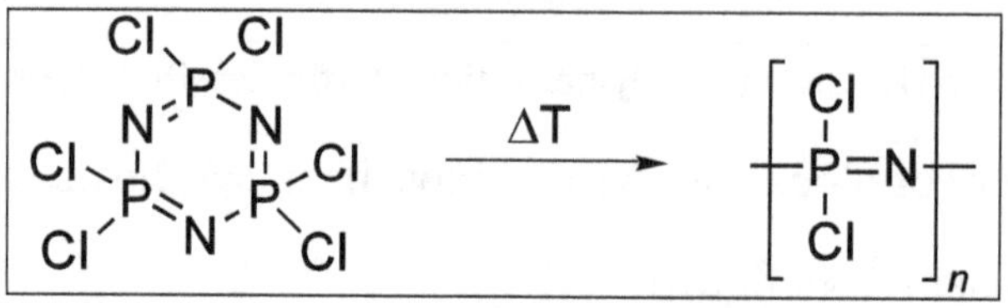

The ROP of hexachloro cyclo-triphosphazene at higher temperatures.

This reaction was first reported at the end of the 19th century by Liebig and Stokes and represents the first example of a synthetic rubber. At higher temperatures the cyclic phosphazene eliminates Cl^- to yield a cyclic cation which then reacts with the free electron pair on a nitrogen of another monomer ring; the P–N single bond in the second ring is broken to yield a linear phosphazene with a $-P(Cl)_2^+$ terminus. The reaction can be catalyzed by Lewis acids such as BCl_3. As the reaction proceeds charge generation along the chain becomes more prevalent and eventually leads to crosslinked insoluble material at conversions much greater than 50%. The poor yields of soluble polymer have proved a major limitation for the commercialization of polyphosphazenes. Thus, alternative, non-ROP, routes have also been investigated, albeit with, as yet, limited commercial success.

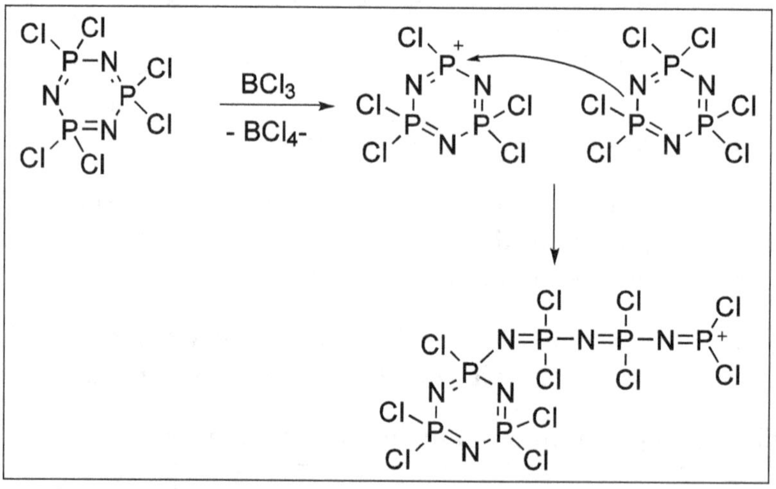

The cationic polymerization of hexachloro cyclo-triphosphazene.

Almost one hundred years after polyphosphazenes were first documented, Allcock's group showed how to convert the base polymer into useful materials.

Since the base material $((PNCl_2)_n)$ is unstable in moist air, all the Cl-atoms must be substituted e.g., by alkoxy or amino groups to convert the so-called "inorganic rubber" into a material which is stable under practical conditions and thus makes the material suitable for a number of applications. Polyphosphazenes with perfluorinated alkoxy side chains have been commercialized under the trade name Eypel F®.

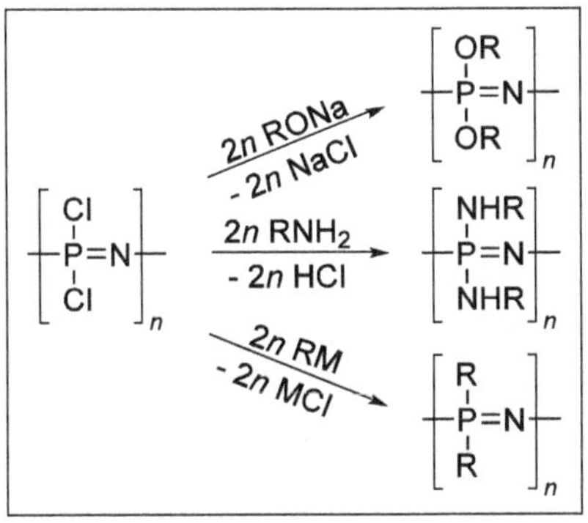

Some substitution reactions to yield useful polyphosphazenes.

By changing the side group substitution the flexibility, refractive index, the transparency (especially for UV light) and the flammability can be adjusted and the hydrophobic or hydrophilic nature of the polymer determined. Phosphazene elastomers having $-OCH_2(CF_2)_nCF_3$ side groups exhibit excellent oil resistance and T_g as low as −60 °C without the addition of plasticizers. They are thus very suitable for special automotive and aerospace applications and have also been suggested for biostable biomedical devices.

Polphosphazenes with oligo-ethylene oxy side chains have also been considered for use in lithium-ion polymer batteries and those with sulfonated aryloxy side groups as membranes for fuel cells. Polyphosphazenes with ethylene oxy side groups can be crosslinked with γ-rays to give hydrogels, which have been investigated as controlled drug release agents. Polyphosphazenes with amino acid side chains are biodegradable and can be employed to aid the regeneration of bone tissue *in vivo*. Polyphosphazenes with aryloxy side groups have also been studied as fire retardants and thermoplastics.

Numerous researchers have tried to substitute the monomer in order to avoid the polymer analogue reactions but substitution of −Cl in the $(PNCl_2)_3$ monomer by $-CH_3$, $-OCH_2CF_3$, −OPh or −Ph leads to molecules which simply decompose or undergo ring expansion on heating. If one of the $-PCl_2$-groups in the ring is exchanged for XCl the ring strain can be increased and a ring-opening polymerization occurs at lower temperature but necessity of subsequent substitution of the −Cl groups and the crosslinking during polymerization remain as disadvantages.

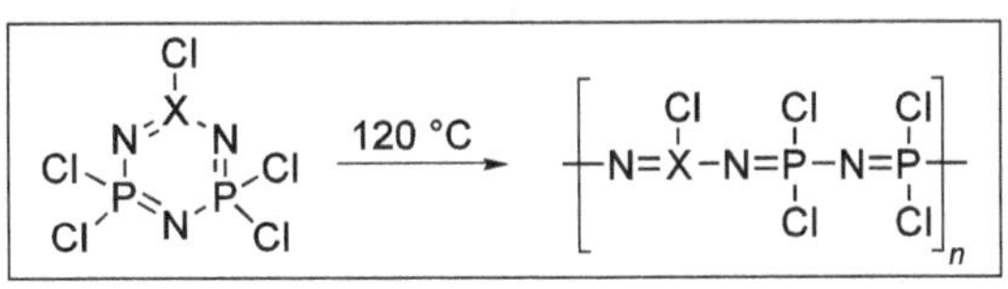

Examples of phosphazene analog monomers with increased ring strain (X = C,S).

An alternative method to increase the ring strain is shown in figure.

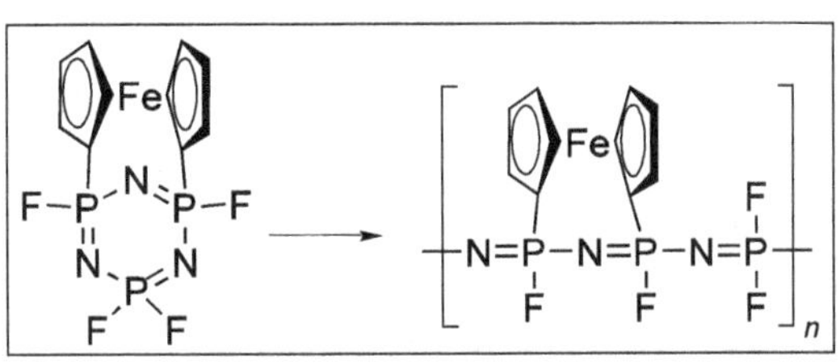

Synthesis of a polyphosphazene with a ferrocene side chain.

Nevertheless, these polymers still require the exchange of the remaining halogens e.g., with $-OCH_2CF_3$ to obtain polymers stable to hydrolysis and this approach has not yet led to a commercial breakthrough for polyphosphazenes.

Free Radical Chain Polymerization

All the monomers from which addition polymers are made are alkenes or functionally substituted alkenes. The most common and thermodynamically favored chemical transformations of alkenes are addition reactions. Many of these addition reactions are known to proceed in a stepwise fashion by way of reactive intermediates, and this is the mechanism followed by most polymerizations. A general diagram illustrating this assembly of linear macromolecules, which supports the name chain growth polymers, is presented here. Since a pi-bond in the monomer is converted to a sigma-bond in the polymer, the polymerization reaction is usually exothermic by 8 to 20 kcal/mol. Indeed, cases of explosively uncontrolled polymerizations have been reported.

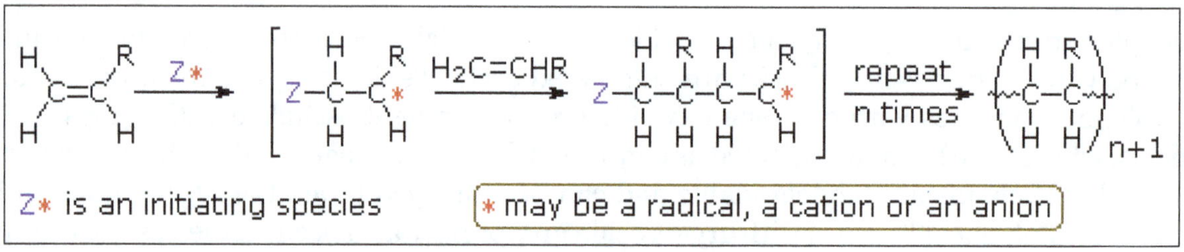

It is useful to distinguish four polymerization procedures fitting this general description.

- Radical Polymerization: The initiator is a radical, and the propagating site of reactivity (*) is a carbon radical.
- Cationic Polymerization: The initiator is an acid, and the propagating site of reactivity (*) is a carbocation.
- Anionic Polymerization: The initiator is a nucleophile, and the propagating site of reactivity (*) is a carbanion.
- Coordination Catalytic Polymerization: The initiator is a transition metal complex, and the propagating site of reactivity (*) is a terminal catalytic complex.

Radical Chain-growth Polymerization

Virtually all of the monomers are subject to radical polymerization. Since this can be initiated by traces of oxygen or other minor impurities, pure samples of these compounds are often "stabilized" by small amounts of radical inhibitors to avoid unwanted reaction. When radical polymerization is desired, it must be started by using a radical initiator, such as a peroxide or certain azo compounds. The formulas of some common initiators, and equations showing the formation of radical species from these initiators are presented below.

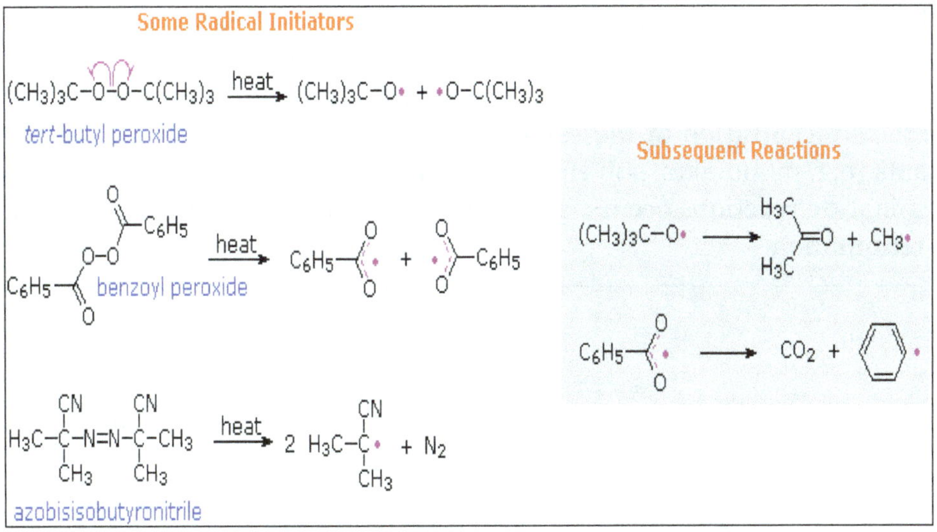

By using small amounts of initiators, a wide variety of monomers can be polymerized. One example of this radical polymerization is the conversion of styrene to polystyrene, shown in the following diagram. The first two equations illustrate the initiation process, and the last two equations are examples of chain propagation. Each monomer unit adds to the growing chain in a manner that generates the most stable radical. Since carbon radicals are stabilized by substituents of many kinds, the preference for head-to-tail regioselectivity in most addition polymerizations is understandable. Because radicals are tolerant of many functional groups and solvents (including water), radical polymerizations are widely used in the chemical industry.

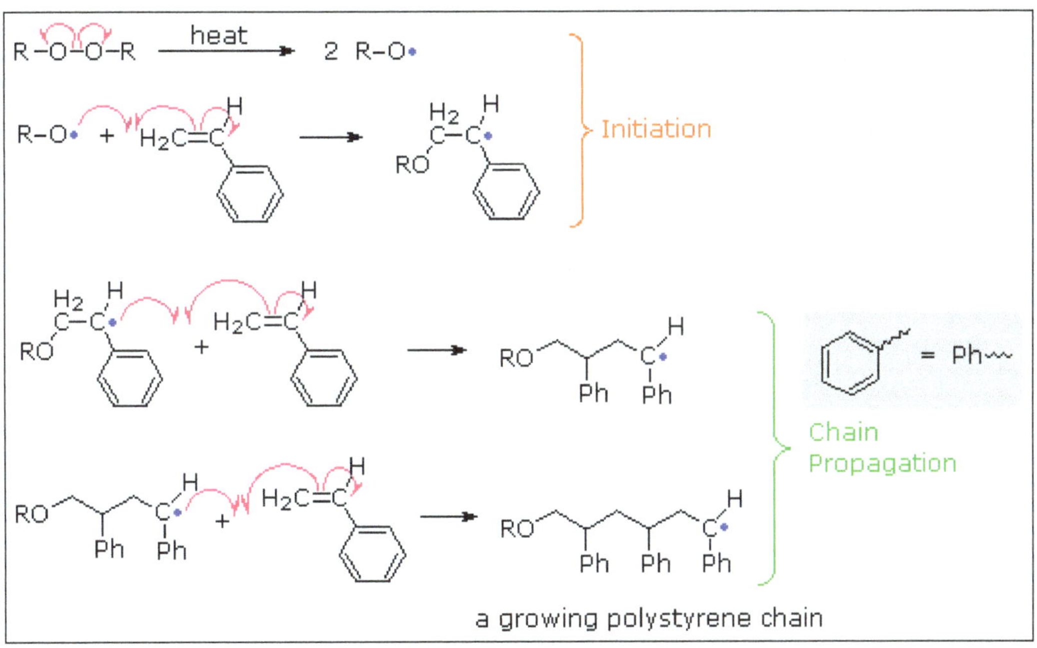

In principle, once started a radical polymerization might be expected to continue unchecked, producing a few extremely long chain polymers. In practice, larger numbers of moderately sized chains are formed, indicating that chain-terminating reactions must be taking place. The most common termination processes are radical combination and disproportionation. These reactions are illustrated by the following equations. The growing polymer chains are colored blue and red, and the hydrogen atom transferred in disproportionation is colored green. Note that in both types of termination two reactive radical sites are removed by simultaneous conversion to stable product(s). Since the concentration of radical species in a polymerization reaction is small relative to other reactants (e.g. monomers, solvents and terminated chains), the rate at which these radical-radical termination reactions occurs is very small, and most growing chains achieve moderate length before termination.

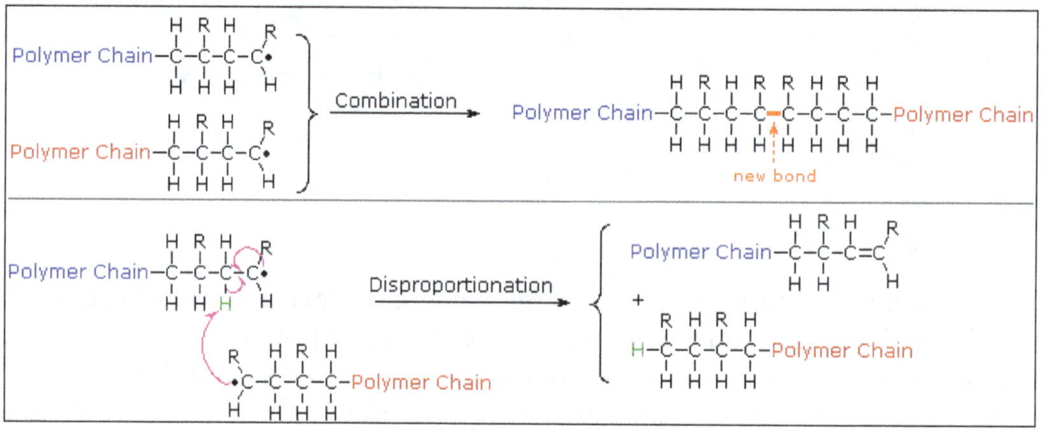

Chain Termination Reactions.

The relative importance of these terminations varies with the nature of the monomer undergoing polymerization. For acrylonitrile and styrene combination is the major process. However, methyl methacrylate and vinyl acetate are terminated chiefly by disproportionation.

Another reaction that diverts radical chain-growth polymerizations from producing linear macromolecules is called chain transfer. As the name implies, this reaction moves a carbon radical from one location to another by an intermolecular or intramolecular hydrogen atom transfer (colored green). These possibilities are demonstrated by the following equations.

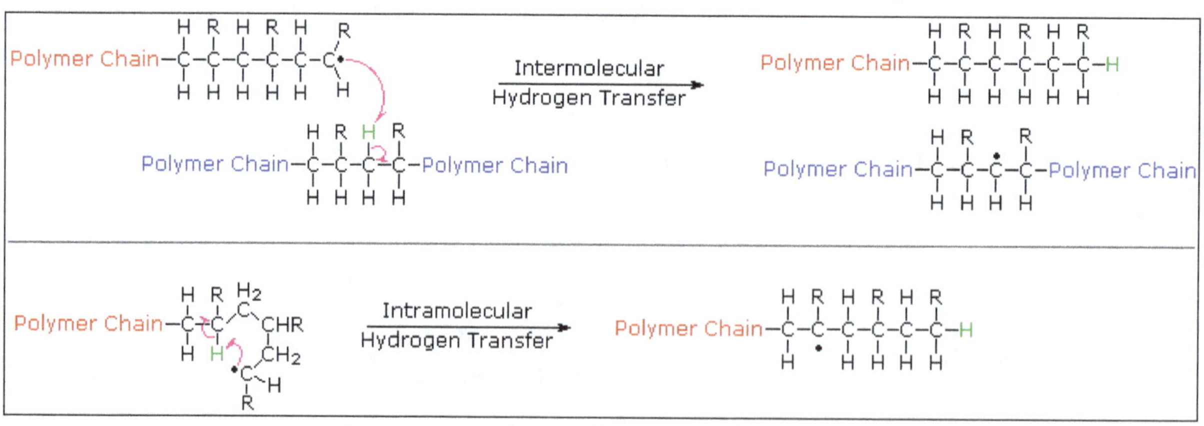

Chain Transfer Reactions.

Chain transfer reactions are especially prevalent in the high pressure radical polymerization of ethylene, which is the method used to make LDPE (low density polyethylene). The 1°-radical at the end of a growing chain is converted to a more stable 2°-radical by hydrogen atom transfer. Further polymerization at the new radical site generates a side chain radical, and this may in turn lead to creation of other side chains by chain transfer reactions. As a result, the morphology of LDPE is an amorphous network of highly branched macromolecules.

Kinetics of Free Radical Chain Polymerization

Radical chain polymerization is a chain reaction consisting of a sequence of three steps—initiation, propagation, and, termination. The initiation step involves two reactions (equations: $I \xrightarrow{k_d} 2R.$ and $R \cdot + M \xrightarrow{k_i} M_1.$). The first is the production of free radicals by homolytic dissociation of an initiator species I to yield a pair of radicals R·.

$$I \xrightarrow{k_d} 2R.$$

where k_d is the rate constant for the catalyst dissociation. The second part of the initiation involves the addition of this radical to the first monomer molecule to produce radical $M_1\cdot$.

$$R \cdot + M \xrightarrow{k_i} M_1.$$

Propagation consists of the growth of $M_n\cdot$ by the successive additions of large numbers (hundreds and perhaps thousands) of monomer molecules. Each addition creates a new radical that has the same identity except that it is larger by one monomer unit. The successive additions may be represented by,

$$M_n \cdot M \xrightarrow{k_d} M_{n+1}\cdot$$

Monomer disappears by the initiation reaction (equation: $I \xrightarrow{k_d} 2R.$) as well as by the propagation (equation: $R \cdot + M \xrightarrow{k_i} M_1.$). The rate of monomer disappearance, also called the rate of polymerization is given by,

$$-\frac{d[M]}{dt} = R_i + R_p$$

where R_i and R_p are the rates of initiation and propagation, respectively. However, the number of monomer molecules reacting in the initiation step is far less than the number in the propagation step for a process producing high polymer. Thus, the rate of initiation can be neglected and the polymerization rate is given simply by the rate of propagation. The rate of propagation, and therefore the rate of polymerization, is the sum of many individual propagation steps. Since the rate constants for all the propagation steps are the same, one can express the polymerization rate by,

$$-\frac{d[M]}{dt} = R_p = k_p[M\cdot][M]$$

where [M] is the monomer concentration and [M·] is the total concentration of all chain radicals, that is, all radicals of size $M_1\cdot$ and larger.

The polymerization rate (equation $-\frac{d[M]}{dt} = R_p = k_p[M\cdot][M]$) is not directly usable because it contains a term of radical concentration. Radical concentrations are difficult to measure quantitatively, since they are very low (~10^{-8} MM), and it is therefore desirable to eliminate [M·] from equation $-\frac{d[M]}{dt} = R_p = k_p[M\cdot][M]$ In order to do this, one can use *steady-state* assumption: the concentration of radicals increases initially, but almost instantaneously reaches a constant, steady-state value. That means the rate of change of the concentration of radicals is zero during the course of the polymerization. Therefore, the rates of initiation R_i and termination R_t of radicals are equal,

$$R_t = R_i = 2k_t[M\cdot]^2$$

Rearrange equation $R_t = R_i = 2k_t[M\cdot]^2$ to obtain,

$$[M\cdot] = \left(\frac{R_i}{2k_t}\right)^{1/2}$$

and substitute equation: $[M\cdot] = \left(\frac{R_i}{2k_t}\right)^{1/2}$ into equation $-\frac{d[M]}{dt} = R_p = k_p[M\cdot][M]$ that yields the rate of polymerization equation.

$$R_p = k_p[M](R_i/2k_t)^{1/2}$$

Rate of Polymerization

The rate of polymerization equation can be modified to include the initiator efficiency (f) as shown in equation $R_p = k_p[M]\sqrt{\frac{fk_d[I]}{k_t}}$. The equation indicates the polymerization rate is proportional to the square root of initiator concentration and to the first power of monomer concentration. Thus, doubling the initiator concentration causes the rate to increase by a factor of about 1.4. This relationship has been confirmed experimentally for a variety of free radical polymerizations. Propagation and termination rate constants, for several commercially important monomers are given in table. The tetrafluoroethylene has an extremely low k_t which results in higher molecular weight polymer as compared with the polymerization of ethylene.

$$R_p = k_p[M]\sqrt{\frac{fk_d[I]}{k_t}}$$

Table: Representative propagation and termination rate constants, k_p and k_t.

Monomer	Temperature (°C)	k_p (L/mol-s)	$k_t \times 10^{-6}$ (L/mol-s)
Acrylonitrile	60	1,960	782

Ethylene	83	240	540
Methyl acrylate	60	2,090	9.5
Methyl methacrylate	60	515	25.5
Styrene	60	176	72
Vinyl acetate	50	2,640	116.8
Vinyl chloride	50	11,000	2,100
Tetrafluoroethylene	40	7,400	7.4×10^{-5}

Average Kinetic Chain Length $\bar{v}$

Another important parameter related to polymerization rate is the average kinetic chain length, $\bar{v}$, which is defined as the average number of monomer units polymerized per chain initiated, which is equal to the rate of polymerization per rate of initiation. Since $R_i = R_t$ under steady-state conditions,

$$\bar{v} = \frac{R_p}{R_i} = \frac{R_p}{R_t}$$

Substituting for R_p and R_t into equation $\bar{v} = \frac{R_p}{R_i} = \frac{R_p}{R_t}$:

$$\bar{v} = \frac{k_p[M][M\cdot]}{2k_t[M\cdot]^2} = \frac{k_p[M]}{2k_t[M\cdot]}$$

Substituting the expression for [M·] from equation $[M\cdot] = \left(\frac{R_i}{2k_t}\right)^{1/2}$,

$$\bar{v} = \frac{k_p[M]}{2(fk_tk_d[I])^{1/2}}$$

Kinetic chain length is related to a variety of rate and concentration parameters. It will decrease as both of initiator concentration and initiator efficiency increase. This is reasonable because increasing the number of growing chains increases the probability of termination. Thus, varying initiator concentration can control molecular weight. In the absence of any side reactions, kinetic chain length is related directly to degree of polymerization depending on the mode of termination. If termination occurs exclusively by disproportionation, $\overline{DP} = \bar{v}$; if it occurs by coupling, $\overline{DP} = 2\bar{v}$.

Autoacceleration, a marked increase in polymerization rate, can occur at very viscous medium even though the chain mobility is reduced and the termination is reduced. The small monomer molecules can still diffuse to the active chain ends to be polymerized. Autoacceleration may cause processing difficulties, particularly in bulk polymerizations, because the increase in rate is usually accompanied by an increase in reaction exotherm and gelation of polymer.

Chain Transfer Reactions

Chain transfer reactions are the process transferring the growing polymer chain to another species which terminates the chain, but at the same time generates a new radical. The chain transfer reactions result in low molecular weight polymer and broad molecular weight distributions. Chain transfer can occur among polymer, monomer, initiator, solvent in the reaction mixture. Equations show the polymer inter- and intrachain transfer reactions, respectively. A chain-end radical may abstract a hydrogen atom from a chain, leading to a reactive site for chain branching.

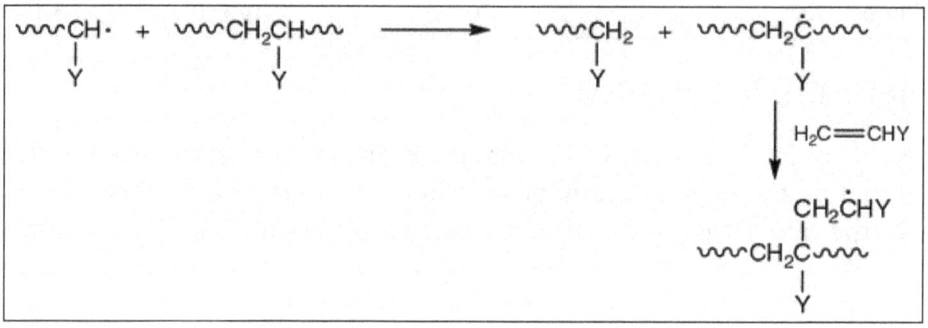

Hydrogen abstraction may also occur intramolecularly, a process referred to as *backbiting*. Polyethylene, produced by high-pressure free radical polymerization of ethylene, is highly branched through backbiting involving five- or six-membered cyclic transition states. The properties of branched polymers are expected to differ remarkably from those of the corresponding linear polymers.

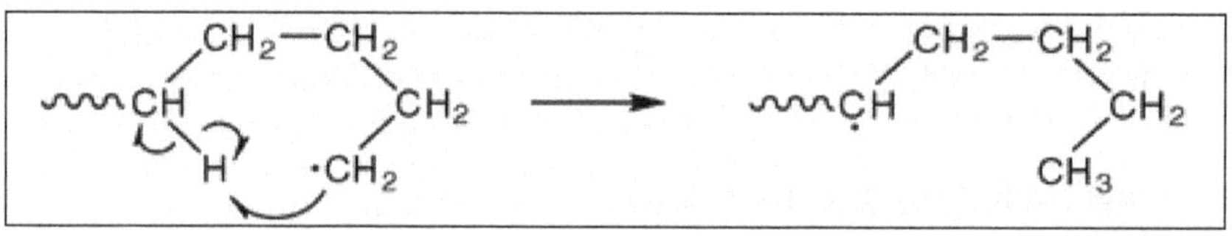

Chain transfer may also occur with initiator or monomer, or it may take place with solvent. Polystyrene prepared in carbon tetrachloride, for example, contains chlorine at the chain end of polymer due to chlorine transfer and initiation by the resultant $\cdot CCl_3$ radicals. Transfer to monomer is particularly important with monomers containing allylic hydrogen, such as propylene, because the formation of resonance-stabilized allylic radicals is highly favorable. Thus, high-molecular-weight polypropylene cannot be prepared using conventional free radical polymerization. The coordination polymerization is used to obtain high molecular weight polypropylene.

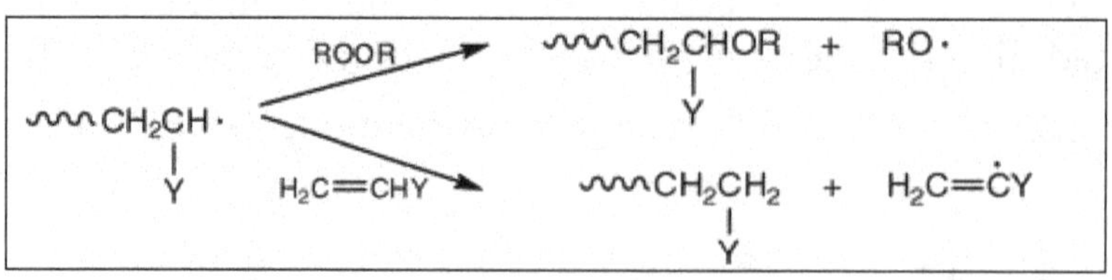

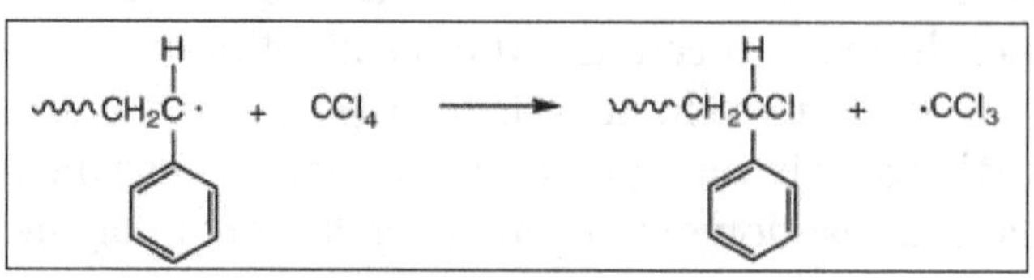

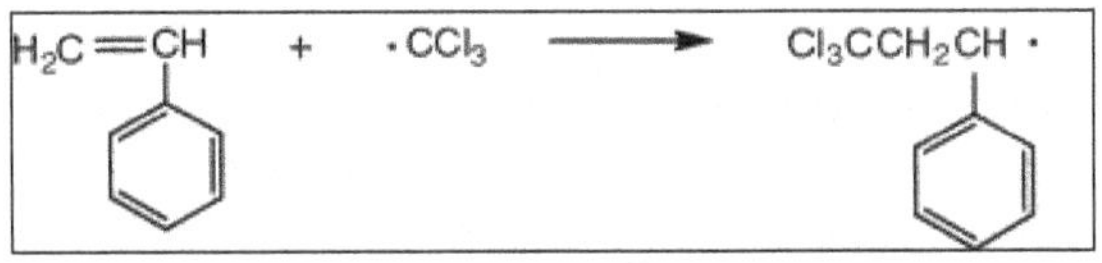

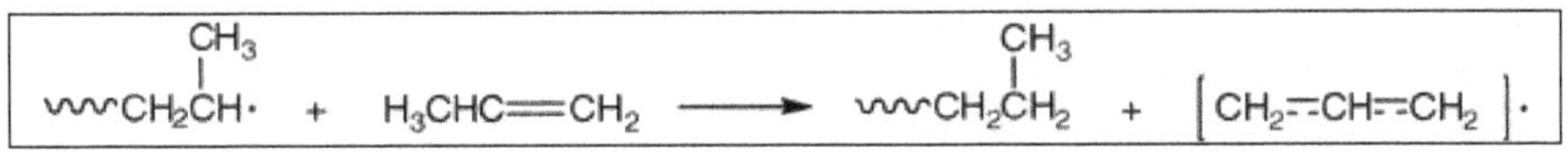

From the molecular weight control point of view, the chain transfer agent plays the most effective way. For instance, a thiol compound has been used widely as chain transfer agent due to its high affinity for hydrogen transfer as:

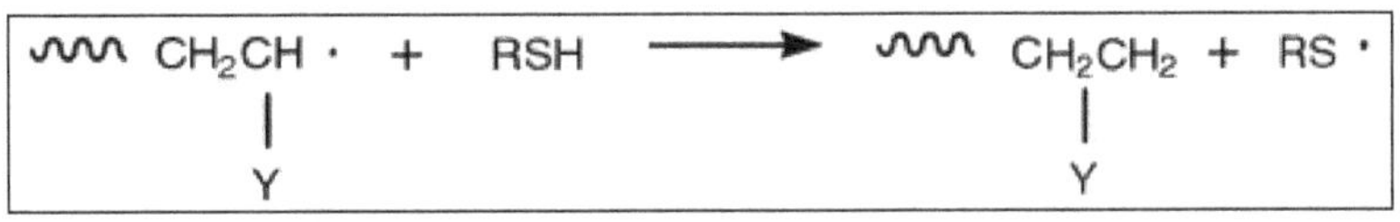

When a chain transfer agent is used in polymerization, it is necessary to redefine the kinetic chain length as being the ratio of propagation rate to the combined rates of termination and transfer, as:

$$\overline{v}_{tr} = \frac{R}{R_i + R_{tr}}$$

Since transfer reactions are second order, with:

$$R_{tr} = k_{tr}[M\cdot][T]$$

Where T is the transfer agent, one can rewrite the expression for $\overline{v}_{tr}$, taking into account all possible transfer reactions as:

$$\overline{v}_{tr} = \frac{k_p[M][M\cdot]}{2k_t[M\cdot]^2 + \sum k_{tr}[M\cdot][T]} = \frac{k_p[M]}{2k_t[M\cdot] + \sum k_{tr}[T]}$$

It is known that:

$$\overline{v} = \frac{k_p[M]}{2k_t[M\cdot]}$$

One can write the reciprocal of equation $\overline{v}_{tr} = \frac{k_p[M][M\cdot]}{2k_t[M\cdot]^2 + \sum k_{tr}[M\cdot][T]} = \frac{k_p[M]}{2k_t[M\cdot] + \sum k_{tr}[T]}$ for $\overline{v}_{tr}$ as:

$$\frac{1}{\overline{v}_{tr}} = \frac{1}{v} + \frac{\sum k_{tr}[T]}{k_p[M]}$$

The ratio of the transfer rate constant to that of propagation is commonly defined as the chain transfer constant, C_T, for a particular monomer:

$$\frac{k_{tr}}{k_p} = C_T$$

Substituting equation $\frac{k_{tr}}{k_p} = C_T$ into equation $\frac{1}{\overline{v}_{tr}} = \frac{1}{v} + \frac{\sum k_{tr}[T]}{k_p[M]}$, one obtains:

$$\frac{1}{\overline{v}_{tr}} = \frac{1}{v} + \frac{\sum C_T[T]}{[M]}$$

As the rate of transfer and concentration of the transfer agent increase, the kinetic chain length becomes progressively smaller. Chain transfer constants for a number of compounds and several of which are given in table.

Table: Representative chain transfer constants C_T, propagation rate constants k_p, and termination rate constants k_t, for styrene and methyl methacrylate (reaction temperature 60 °C).

Transfer Agent	$C_T \times 10^4$	
	Styrene	Methyl Methacrylate
Benzene	0.023	0.04
Toluene	0.125	0.20
Chloroform	0.5	1.77
Carbon tetrachloride	90	2.40
Carbon tetrabromide	22,000	2,700
1-Butanethiol	210,000	6,600
k_p (L/mol-s)	176	515
$k_t \times 10^{-6}$ (L/mol-s)	73	25.5

When the concentration of transfer agent is high and k_{tr} is much greater than k_p, very low molecular weight polymers called telomers are obtained. The process is called telomerization. Chain transfer reactions can also be used to prevent free radical polymerizations. One type of compound, added as a stabilizer to vinyl monomers, is an alkylated phenol 10, which can transfer its phenolic hydrogen to form a new radical that undergoes coupling reaction rather than initiating polymerization. Such compounds, called inhibitors, are commonly added to monomers to prevent premature polymerization during shipment or storage.

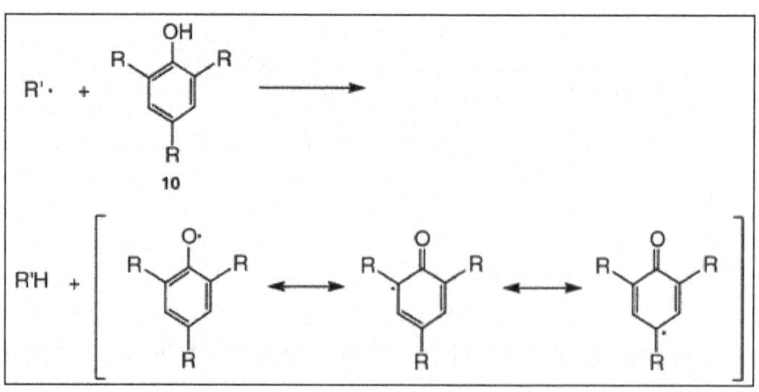

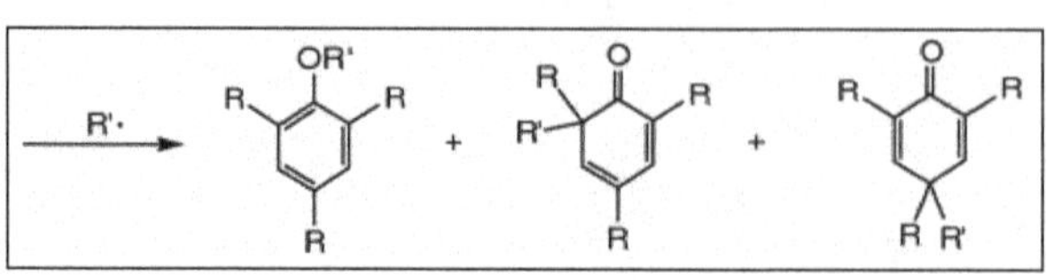

Living Polymerization

When the termination reaction is not present in the polymerization of monomers, we call the polymerization is living polymerization. An important consequence of living polymerization is that the average degree of polymerization is simply equal to the ratio of the initial monomer concentration to the initiator concentration as shown in equation. $\overline{DP} = \frac{[M]_0}{[I]_0}$ There are different approaches to achieve living radical polymerizations:

$$\overline{DP} = \frac{[M]_0}{[I]_0}$$

Living Radical Polymerization

The living radical polymerization is also called controlled radical polymerization. The formal IUPAC name of this type of polymerization is called reversible deactivation radical polymerization (RDRP). The lifetime in the conventional radical polymerization is very short (in second) because of the presence of bimolecular termination. The living radical polymerization can be achieved by minimizing normal bimolecular termination and prolonging the lifetime of living polymers. Special mode of reaction for the propagating radicals by either reversible termination or reversible transfer has been developed. Living radical polymerization (LRP) with reversible termination generally proceeds as follows:

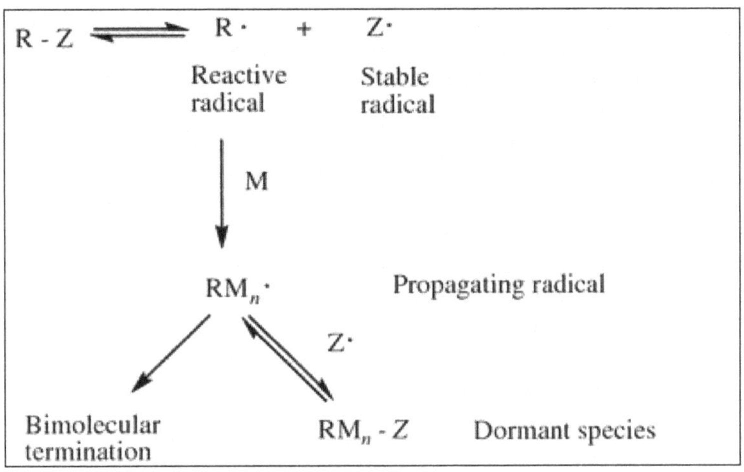

The initiator RZ undergoes homolytic bond breakage to produce one reactive and one stable free radical. The reactive radicals quickly initiate polymerization, but the stable radicals are too stable to initiate polymerization. However, the stable radicals preserve the propagation chain being living without any termination. For living polymerization, it is important to have all the initiator decomposes at once so that all propagating radicals grow almost at the same time. Fast initiation is important, but it is the fast equilibrium between the propagating radical and dormant species with an appropriate equilibrium constant that determines the living characteristics of polymerization. The equilibrium constant must be low but not too low; that is, the concentration of propagating radical must be sufficient to achieve a reasonable propagation rate but not so high that normal bimolecular termination becomes important.

At the beginning of the reaction, the concentrations of propagating and stable radicals are equal, but they change rapidly with the progress of reactions. The concentrations of stable radicals are increasing while the concentrations of propagating radicals are decreasing because the reactions are shifted toward more stable species as shown in equation. Overall, the concentration of stable radicals is about four orders higher than that of propagating radicals. The concentration of propagating radicals is about the same or lower than the conventional radical polymerization. The stable radicals function as controlling agent to form reversible dormant species with propagating radicals. The equilibrium favors the dormant species by several orders of magnitude as compared with the propagating radicals. Thus, the concentration of dormant species is about six orders higher than that of propagating radical. In short, the introduction of the dormant species in the radical polymerization suppresses the bimolecular termination reaction to have a living polymer. The average life time of the living polymer has been increased by at least four orders of magnitude. Thus, the living free radical polymerization becomes possible due to the formation of stable radicals and reversible dormant species. By sequential living polymerization of one kind of monomer to another kind of monomer, well-defined block copolymers can be obtained through living polymerization. The second type of monomers usually has to add quickly after the first kind monomers are consumed to avoid any bimolecular reactions among propagating radicals. Currently, reversible termination reactions and reversible chain transfer reactions are employed in the living free radical polymerizations.

Atom Transfer Radical Polymerization

The reactions of atom transfer radical polymerization (ATRP), involves an atom transfer, usually a halogen atom, to form propagating radicals and dormant species from the complex of atom transfer agent and copper (I) catalyst. For instance, the polymerization of styrene using 1-chloro-1-phenylethane (11, or the bromo analog) as initiator in the presence of a copper (I) bipyridyl (bpy) complex. Initiation occurs when a halogen atom is transferred from 11 to the complex. The resultant 1-phenylethyl radical in turn adds to a styrene molecule. The halogen atom is then reversibly transferred to the styryl radical, thus preventing radical termination reactions from occurring while allowing the propagation reaction to proceed.

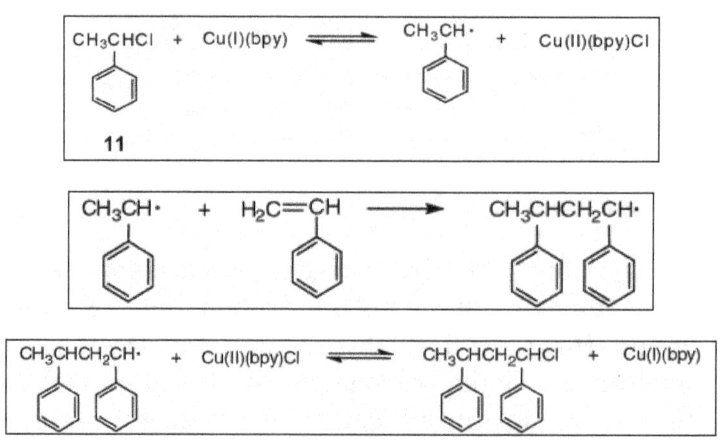

Nitroxide-mediated Polymerization (NMP)

Stable nitroxide radical such as TEMPO 12 can mediate polymerization as shown in equation. The TEMPO is too stable to initiate polymerization, but it does promote decomposition of benzoyl peroxide to benzoyloxy radicals which initiate chain growth. Then TEMPO combines reversibly with the growing polymer chain ends which protect the chain end radicals from termination. Dissociation in turn frees the chain end radical to add to other styrene molecules.

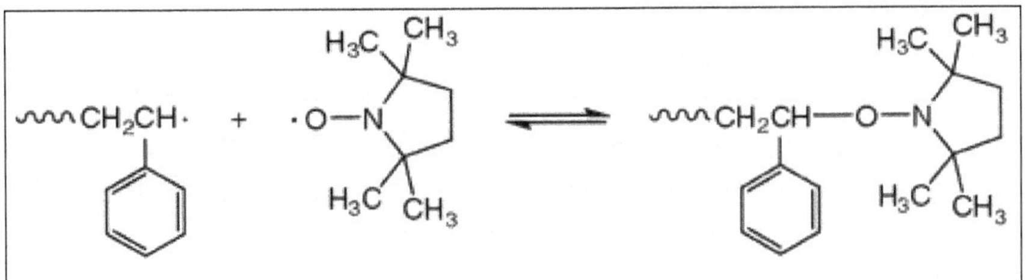

12

Radical Addition-Fragmentation Transfer (RAFT)

ATRP and NMP control chain growth by reversible termination. RAFT living polymerizations control chain growth through reversible chain transfer. A chain-transfer agent R'CSSR, such as cumyl dithiobenzoate 13 reversibly transfer a labile end group (a dithioester end group) to a propagating chain. The polymerization is carried out with a conventional initiator such as a peroxide or AIBN in the presence of the chain-transfer agent. The key that makes RAFT a living polymerization is the choice of the RAFT transfer agent. Living polymerization occurs with dithioesters because the transferred end group in the polymeric dithioester is as labile as the dithioester group in R'CSSR. This results in an equilibrium between dormant polymer chains and propagating radicals (equation with K = 1), which controls the living polymerization.

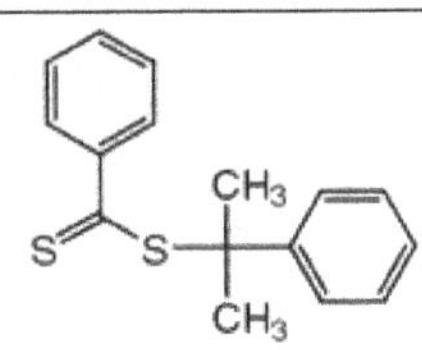

13

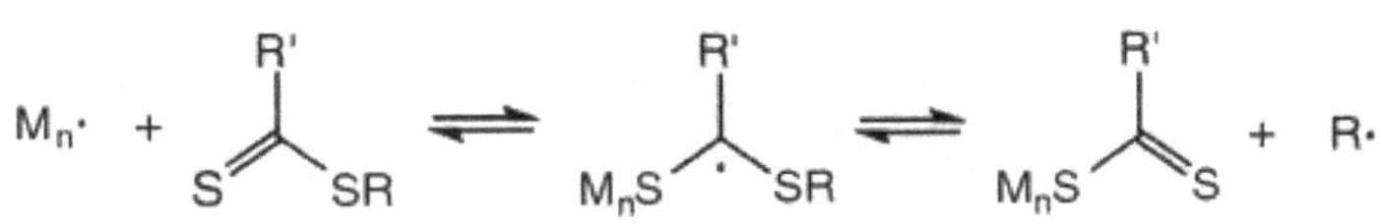

RAFT works with a wider range of monomers than TEMPO and ATRP. RAFT does not produce polymers with metal catalysts but produce polymers with dithioester group with odors and colors. The design and synthesis of RAFT agent has been thoroughly reviewed and outlined. The methods are extended to functional RAFT agent and macro-RAFT agent.

The limitations of living radical polymerization are that irreversible bimolecular termination of propagating radicals will occur at high monomer conversion, polyfunctional initiators, high initiator concentration, and high target molecular weight (>100,000).

Polymerization of Dienes

Polymerization of isolated dienes leads to crosslinked polymers when the double bonds react independently of each other. In certain cases, the addition reactions can occur within the double bonds of molecule and lead to cyclic polymers. Polymerization of divinylformal 14 illustrates this process as shown in equation.

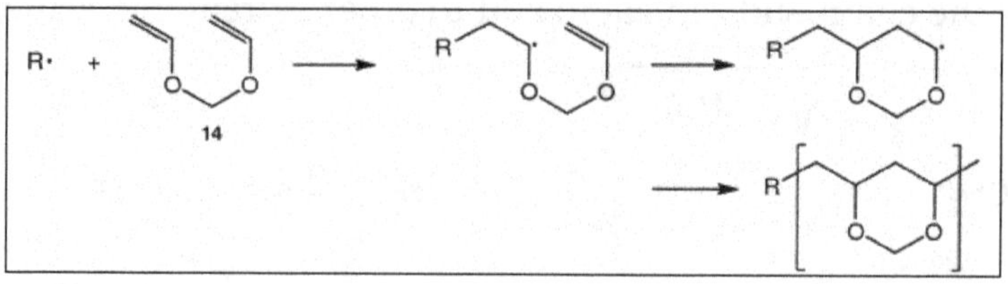

The normal (head-to-tail) mode of addition yields the six-membered ring, whereas abnormal (head-to-head) addition gives the five-membered ring. The former is preferred, but both are formed in significant amounts.

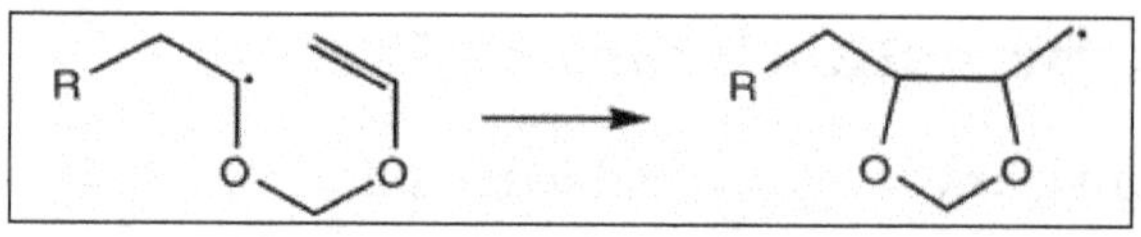

Several diallyl monomers such as 15 and 16 are used commercially to prepare highly cross-linked thermosetting allyl resins or for crosslinking other polymers. Diallyl phthalate 15 is used widely in the manufacturing of printed wiring boards, electrical insulations, fiber reinforced composites, etc. Diethylene glycol bis (allyl carbonate) 16 is used to make optical clear polymers for the applications in eyewares, lenses, camera filters, panel covers, and the like.

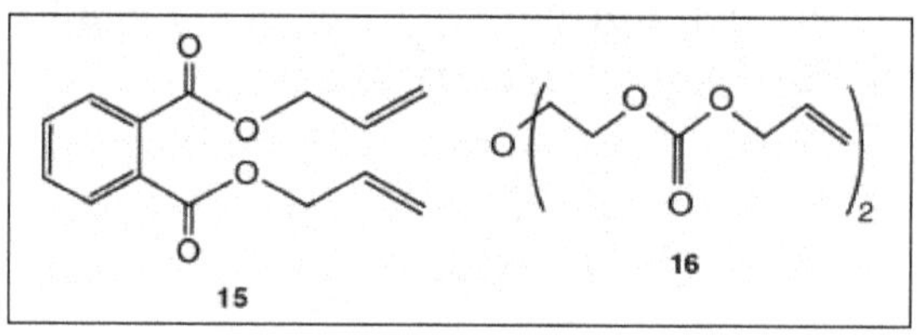

Conjugated dienes such as 1,3-butadiene 17 undergo both 1,2- and 1,4-addition to obtain polymers through the delocalized radical intermediate 18. Thus 1,2-addition gives polymer 19 with pendant vinyl groups, while 1,4-addition leads to polymer with unsaturation in the chain. In the latter case, both cis 20 and trans 21 configurations are possible.

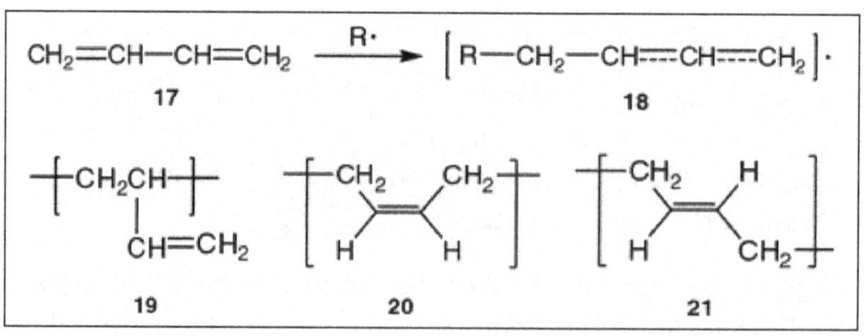

With substituted dienes like isoprene (2-methyl-1, 3-butadiene) the situation is more complicated; 1, 2 and 3,4 structures (22 and 23, respectively) as well as cis- and trans-1, 4 (24 and 25), are possible. All are formed in free radical polymerization but as expected the head-to-tail trans-1, 4 25 predominates. Natural Hevea rubber contains the head-to-tail of 1, 4-polyisoprene units with cis structure 24.

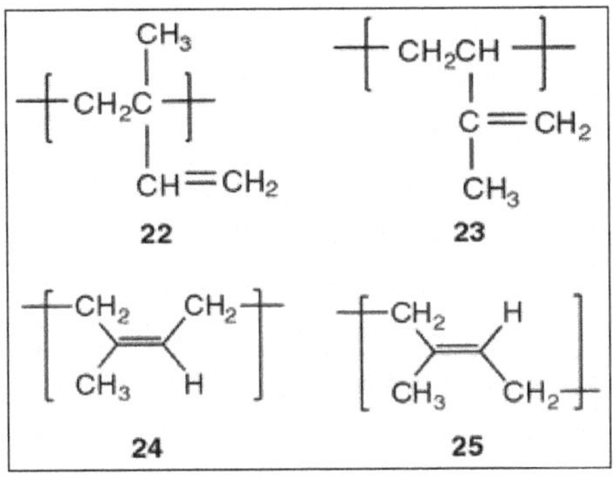

Table lists the percentages of each structural unit at different polymerization temperatures for the three most important diene monomers—butadiene, isoprene, and chloroprene (2-chloro-1,3-butadiene). It is interesting to note that, for isoprene, the amount of cis-1,4 structure increases with increasing temperature up to about 100 °C, then it decreases. The total amount of 1,2 and 3,4 polymer (approximately 10 %) does not change appreciably with temperature; however, the 3,4 structure is favored at higher temperatures. With chloroprene, the trans-1,4 structure also decreases with increasing temperature, but the amount of trans-1,4 is slightly higher than that of isoprene over the temperature range measured. The results may be from the larger size of Cl group than that of CH3 group, a steric hinderance dominated factor.

Table: Compositions of free radical-initiated diene polymers.

Monomer	Polymerization temperature(°C)	Percent			
		cis-1,4	trans-1,4	1,2	3,4
Butadiene	−20	6	77	17	–
	20	22	58	20	–
	100	28	51	21	–
	233	43	39	18	–
Isoprene	−20	1	90	5	4
	−5	7	82	5	5
	50	18	72	5	5
	100	23	66	5	6
	257	12	77	2	9
Chloroprene	−46	5	94	1	0.3
	46	10	81–86	2	1
	100	13	71	2.4	2.4

Temperature Effect of the Free Radical Polymerization

By varying the polymerization temperature, there is big effect on the rate and degree of polymerization. When the temperature increases, the rate of polymerization increases, but the degree of polymerization decreases as shown in figure. The quantitative effect of temperature is complicated.

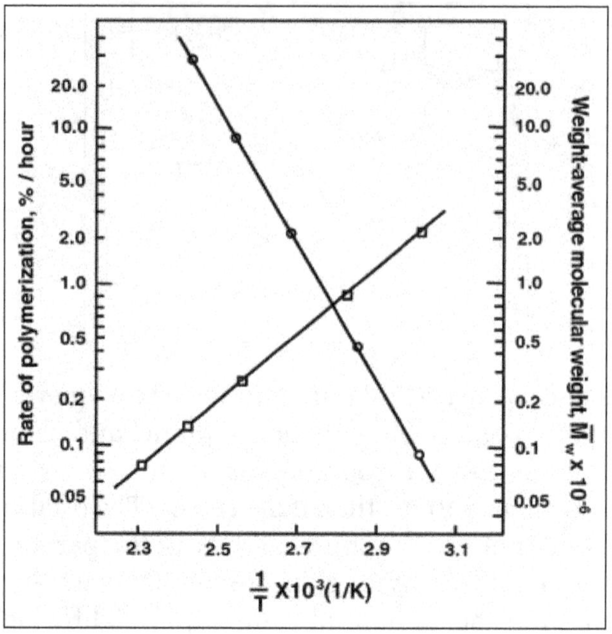

Dependence of the polymerization rate (○) and polymer molecular weight (□) on the temperature for the thermal self-initiated polymerization of styrene.

Activation Energy and Frequency Factor

The quantitative effect of temperature on rate of polymerization (R_p) and degree of polymerization $\left(\overline{X}_n\right)$ is very complex since they depend on a combination of three rate constants of initiation (kd), propagation (k_p) and termination (k_t). Each of the rate constants can be expressed by an Arrhenius equation,

$$k = Ae^{-E/RT}$$

Or

$$\ln k = \ln A - \frac{E}{RT}$$

where A is the collision frequency factor, E the Arrhenius activation energy, and T the kelvin temperature. A plot of ln k versus 1/T allows the determination of both E and A from the slope and intercept, respectively. Table shows the values of the frequency factor and activation energy of propagation (E_p) and termination (E_t) of several monomers.

Table: Kinetic parameters in radical chain polymerization.

ns-polymerization to produce Synth	E_p	$A_p \times 10^{-7}$	E_t	$A_t \times 10^{-9}$
1,3-Butadiene	24.3	12	–	–
2-Vinylpyridine	33	–	21	–
Acrylonitrile	16.2	–	15.5	–
Ethylene	18.4	–	1.3	–

Methyl acrylate	29.7	10	22.2	15
Methyl methacrylate	26.4	0.087	11.9	0.11
Styrene	26	0.45	8.0	0.058
Vinyl acetate	18	3.2	21.9	3.7
Vinyl chloride (50 °C)	16	0.33	17.6	600
Tetrafluoroethylene(83 °C)	17.4	–	–	–

E_p values are in kJ/mole of polymerizing monomer; E_t values are in kJ/mole of propagating radicals; A_p and A_t values are in liters/mole-second. The data are obtained at 60 °C unless otherwise noted.

The variations in the values of the frequency factor of propagation (A_p) are greater than those in E_p which indicates that steric effects are probably the more important factor to determine the absolute value of k_p. Thus, the more hindered monomers such as methyl methacrylate have lower k_p and A_p values than the less hindered ones such as methyl acrylate. The A_p values in general are lower than the usual value (10^{11}–10^{13}) of the frequency factor of a bimolecular reaction. The result is probably due to a large decrease in entropy on polymerization. The variations in the values of the frequency factor of termination (A_t), generally follow the trend of A_p values, but with larger values.

Rate of Polymerization

For a polymerization initiated by the thermal decomposition of an initiator, the polymerization rate depends on the ratio of three rate constants $k_p(k_d/k_t)^{1/2}$ according to equation. $R_p = k_p[M]\sqrt{\frac{fk_d[I]}{k_t}}$

The temperature dependence of this ratio can be obtained by combining three separate Arrhenius equations as below.

$$In\left[k_p\left(\frac{k_d}{k_t}\right)^{1/2}\right] = In\left[A_p\left(\frac{A_d}{A_t}\right)^{1/2}\right] - \frac{\left[E_p + \left(\frac{E_d}{2}\right) - \left(\frac{E_t}{2}\right)\right]}{RT}$$

The overall activation energy of rate of polymerization E_R is equal to,

$E_p + (E_d/2) - (E_t/2)$. From equation $R_p = k_p[M]\sqrt{\frac{fk_d[I]}{k_t}}$, one can rewrite the above equation as,

$$In\ R_p = In\left[Ap\left(\frac{A_d}{A_t}\right)^{1/2}\right] + In\left[(f[I])^{\frac{1}{2}}[M]\right] - \frac{E_R}{RT}$$

By plotting ln R_p versus 1/T, one can obtain E_R and $A_p(A_d/A_t)^{1/2}$ from the slope and intercept, respectively.

Table shows the thermal properties of commonly used initiators. The activation energy of initiator decomposition is in the range of 120–150 kJ mole^{-1} in general. The E_p and E_t value of most monomers are in the ranges of 20–40 kJ mole^{-1} and 8–20 kJ mole^{-1} respectively. Thus the ER of most polymerization initiated by thermal initiator decomposition is about 80–90 kJ mole^{-1}. This corresponds to a two to three time rate increase for every 10°C temperature increase. The situation is different for other mode of initiation. The redox initiation (e.g., Fe^{2+} with thiosulfate or cumene hydroperoxide) usually takes place at lower temperature as compared to the thermal polymerizations because the former has lower activation energy. The E_d value of redox initiation is only about 40–60 kJ mole^{-1} or about 80 kJ mole^{-1} less than that of thermal initiation. This leads to an ER redox polymerization about 40 kJ mole^{-1} or about half of the nonredox initiators.

Table: Properties of common thermal initiators.

Initiator	$k_d \times 10^5$	T (°C)	E_d
2,2′-Azobisisobutyronitrile	0.845	60	123.4
Acetyl peroxide	2.39	70	136.0
Benzoyl peroxide	5.50	85	124.3
Cumyl peroxide	1.56	115	170.3
t-Butyl peroxide	3.00	130	146.9
t-Butyl hydroperoxide	0.429	155	170.7

- All data are for decompositions in benzene solution.
- The units of k_d are s^{-1}; the units of E_d are kJ/mole.

The initiation step of photo polymerization is temperature independent (E_d = 0), because the decomposition of initiator is induced by light rather than by heat. The overall activation energy of photo polymerization is then only about 20 kJ mole^{-1} which indicates this polymerization is relatively insensitive to temperature. However, most photo initiators can also be decomposed thermally; the initiators may undergo appreciable thermal decomposition in addition to photo decomposition at higher temperatures. In such cases, one must take into account both the thermal and photo initiations. The initiation and overall activation energies for a purely thermal self-initiated polymerization are approximately the same as for initiation by the thermal decomposition of an initiator. For the thermal, self-initiated polymerization of styrene, its E_d is 121 kJ mole^{-1} and E_R is 86 kJ mole^{-1}. However, pure thermal polymerization proceeds at very slow rates because of very low frequency factor (104–106) resulting in very low probability of initiation process.

Degree of Polymerization

The effect of temperature on the molecular weight of the polymer produced by thermally initiated polymerization will be determined by the ratio of $k_p/(k_d/k_t)^{1/2}$ according to equation. $\bar{v} = \dfrac{k_p[M]}{2(fk_t k_d[I])^{1/2}}$ of the degree of polymerization, if transfer reactions are negligible. The variation of the ratio with temperature can be expressed by:

$$\ln\left[\frac{k_p}{(k_d k_t)^{1/2}}\right]=\ln\left[\frac{A_p}{(A_d A_t)^{1/2}}\right]-\frac{\left[E_p-\left(\frac{E_d}{2}\right)-\left(\frac{E_t}{2}\right)\right]}{RT}$$

The overall activation energy of degree of polymerization $\left(E_{\overline{X}_n}\right)$ is equal to $\left[E_p-(E_d/2)-(E_t/2)\right]$. Then the temperature effect on the degree of polymerization can be expressed by,

$$\ln\overline{X}_n=\ln\left[\frac{A_p}{(A_d A_t)^{1/2}}\right]+\ln\left[\frac{[M]}{(f[I])^{1/2}}\right]-\frac{E_{\overline{X}_n}}{RT}$$

$E_{\overline{X}_n}$ has a value of about −60 kJ mole^{-1} and $\overline{X}_n$ decreases rapidly with increasing temperature. $E_{\overline{X}_n}$ is about the same for a purely thermal, self-initiated polymerization. For a pure photo polymerization, $E_{\overline{X}_n}$ is positive by approximately 20 kJ mole^{-1}, since E_d is zero, and $\overline{X}_n$ increases moderately with temperature. For all other cases, $\overline{X}_n$ decreases with temperature.

When chain transfer occurs in the polymerization, the temperature effect on the $\overline{X}_n$ depends on the relative importance of the chain transfer reaction. For the case where chain transfer to chain transfer agent T, one can obtain the following equation according to equation. $\frac{1}{\overline{v}_{tr}}=\frac{1}{\overline{v}}+\frac{\sum C_T[T]}{[M]}$

$$-\ln\left[\frac{[M]}{[T]}\left(\frac{1}{\overline{X}_n}-\frac{1}{(\overline{X}_n)_0}\right)\right]=\ln\frac{k_p}{k_{tr}}=\ln\frac{A_p}{A_{tr}}-\frac{(E_p-E_{tr})}{RT}$$

Now, $E_{\overline{X}_n}$ is equal to (E_p-E_{tr}) and can be obtained by plotting the left side of equation versus 1/T. The (E_p-E_{tr}) is usually in the range of −20 to −65 kJ more^{-1} and the molecular weight decreases with increasing temperature. Table shows the activation parameters for chain transfer in styrene polymerization using different chain transfer agents. The frequency factors for transfer reactions are usually greater than those of propagations.

Table: Activation parameters for chain transfer in styrene polymerization (60 °C).

Transfer agent	$-(E_p-E_{tr,S})$	$\log(A_{tr,S}/A_p)$
Benzene	62.0	3.9
Carbon tetrachloride	20.9	1
Cyclohexane	56.1	3.1
Ethyl benzene	23.0	−0.55
Isopropyl benzene	23.0	−0.47
n-Butyl bromide	46.1	2

n-Butyl chloride	58.6	4
n-Butyl iodide	29.3	1
t-Butyl benzene	57.4	3.8
Toluene	42.3	1.7

Thermodynamics of Free Radical Polymerization

The thermodynamics of free radical polymerization can be expressed by $\Delta G_p = \Delta H_p - T\Delta S_p$. The ΔG_p, ΔH_p, and ΔS_p for a polymerization are the difference in free energy, enthalpy, and entropy, respectively, between 1 mol of monomer and 1 mol of repeating units in the polymer product. The thermodynamic properties of a polymerization relate only to the propagation step, since polymerization consists of single acts of initiation and termination and a large number of propagation steps.

Chain polymerizations of alkene are exothermic (ΔH_p is "negative") and exoentropic (ΔS_p is "negative"). The exothermic nature of polymerization arises because the process involves the exothermic conversion of π-bonds (less stable) in monomer molecules into σ-bond (more stable) in the polymer. The negative ΔS_p for polymerization arises from the decreased degrees of freedom for the polymer relative to the monomer. Thus, polymerization is favorable from the enthalpy viewpoint, but unfavorable from the entropy viewpoint. Table shows the wide range of ΔH_p values for various monomers and also shows the general thermodynamic feasibility for any carbon–carbon double bond (ΔG_p is negative). However, thermodynamic feasibility does not indicate the experimental conditions of polymerization. The ΔS_p values fall in a narrower range of values. ΔS_p is in a small range ~100 to ~150 J/K-mol and ΔG_p is "negative" due to $-T\Delta S_p$ over weight ΔH_p. Thus, polymerization is a favored reaction from thermodynamic point of view.

Table: Enthalpy and entropy of polymerization at 25 °C.

Monomer	−ΔH	−ΔS
Ethylene	93	155
Propene	84	116
1-Butene	83.5	113
Isobutylene	48	121
1,3-Butadiene	73	89
Isoprene	75	101
Styrene	73	104
α-Methylstyrene	35	110
Vinyl chloride	72	–
Vinylidene chloride	73	89
Tetrafluoro ethylene	163	112
Acrylic acid	67	–
Acrylonitrile	76.5	109
Maleic anhydride	59	–
Vinyl acetate	88	110
Methyl acrylate	78	–
Methyl methacrylate	56	117

ΔH refer to the conversion of liquid monomer to amorphous or (slightly) crystalline polymer. ΔS refers to the conversion of monomer (at a concentration of 1 M) to amorphous or slightly crystalline polymer. The subscripts lc are often used with ΔH and ΔS to show the initial and final states (that is, ΔH_{lc} and ΔS_{lc}). The units of ΔH are kJ/mole of polymerized monomer; the units of ΔS are J/K-mole

Monomer Reactivity

Although the polymerization reaction is thermodynamically favored, the kinetic feasibility of polymerization varies considerably from one monomer to another depending on the stability of monomer radical and polymer radical.

As shown in table, the ΔH values for ethylene, propene, and 1-butene are very close to the difference (82–90 kJ/mole) between the bond energies of the π-bond in an alkene and the σ-bond in an alkane. The ΔH values for the other monomers vary considerably. The variations in ΔH for differently substituted ethylenes arise from any of the following effects: (1) differences in the resonance stabilization of monomer and polymer due to differences in conjugation or hyperconjugation, (2) steric strain differences in the monomer and polymer arising from bond angle deformation, bond stretching, or interactions between nonbonded atoms, and (3) differences in hydrogen bonding or dipole interactions in the monomer and polymer. The following monomers (I) can be stabilized through resonance that decreases ΔH of polymerization. However, the monomer (II) is poorly conjugated and its ΔH is similar to that of ethylene.

Type I Monomer

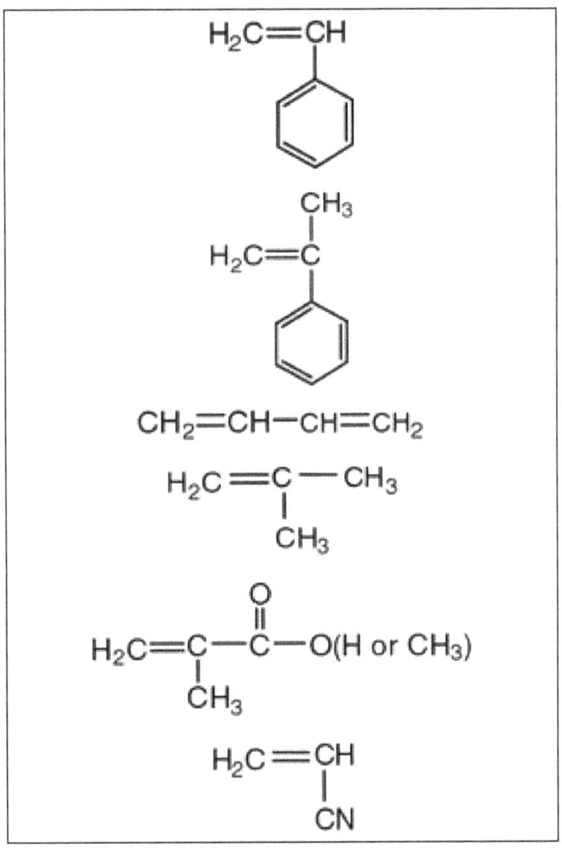

Type II Monomer

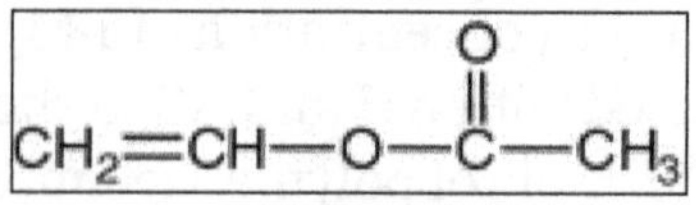

Many substituents stabilize the monomer but have no appreciable effect on polymer stability, since resonance is only possible with the former. The net effect is to decrease the exothermicity of the polymerization. Thus, hyperconjugation of alkyl groups with the C = C lowers ΔH for propylene and 1-butene polymerizations. Conjugation of the C = C with substituents such as the benzene ring (styrene and α-methylstyrene), and alkene double bond (butadiene and isoprene), the carbonyl linkage (acrylic acid, methyl acrylate, methyl methacrylate), and the nitrile group (acrylonitrile) similarly leads to stabilization of the monomer and decreases enthalpies of polymerization. When the substituent is poorly conjugating as in vinyl acetate, the ΔH is close to the value for ethylene.

The effect of 1,1-disubstitution manifests itself by decreased ΔH values. This is a consequence of steric strain in the polymer due to interactions between substituents on alternating carbon atoms of the polymer chain. In 25, the polymer chain is drawn in the plane of the text with the H and Y substituents placed above and below the plane of the text. The dotted and triangular lines indicate substituents below and above this plane, respectively. Such interactions are referred to as 1,3-interactions and are responsible for the decreased ΔH values in monomers such as isobutylene, α-methyl styrene, methyl methacrylate, and vinyl chloride. The effect in α-methyl styrene is especially significant. The ΔH value of −35 kJ/mole is the smallest heat of polymerization of any monomer. To summarize, the stability of the polymer radical is more important than the stability of the monomer toward addition of a free radical in the propagation reactions.

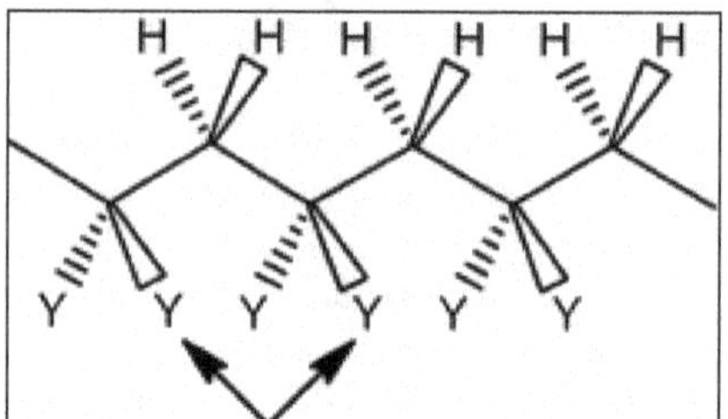

1,3-Interaction in polymer

25

Ceiling Temperature

The reversibility of vinyl polymerization reactions has been observed which depends on the monomer reactivity and the stability of polymer radical as shown in the following:

$$M_x \cdot + M \underset{k_{dp}}{\overset{k_p}{\rightleftarrows}} M_{(x+1)} \cdot$$

Where k_{dp} is the depropagation rate constant. Just as radical stability retards propagation, it enhances depropagation. As the polymerization temperature is raised, the depropagation rate increases until a point is reached where the forward and back reactions are equal. The temperature at which this occurs is called the ceiling temperature (T_c), and at that temperature ΔG of polymerization is zero, or:

$$T_c = \frac{\Delta H}{\Delta S}$$

It follows that whatever affects ΔH or ΔS for a given monomer (e.g., monomer concentration, pressure) will also affect T_c. Table lists ceiling temperatures for several pure liquid monomers.

Table: Representative ceiling temperatures, T_c, of pure liquid monomers.

Monomer	T_c (°C)
1,3-Butadiene	585
Ethylene	610
Isobutylene	175
Isoprene	466
Methyl methacrylate	198
α-Methyl styrene	66
Styrene	395
Tetrafluoro ethylene	1,100

Characteristics of ΔS Values of Free Radical Polymerization

While the ΔH values vary over a wide range for different monomers, the ΔS values are less sensitive to monomer structure, being relatively constant within the range of ~100–120 J/K-mole. The TΔS contribution to the ΔG of polymerization will be small and will vary only within a narrow range. Thus, the variation in the TΔS term at 50°C for all monomers is in the narrow range of 30–40 kJ/mole. The ΔS of polymerization arises primarily from the loss of the translation entropy of the monomer. Losses in the rotational and vibration entropies of the monomers are essentially balanced by gains in the rotation and vibration entropies of the polymer. Thus, ΔS for polymerization is essentially the translational entropy of the monomer, which is relatively insensitive to the structure of the monomer.

Molecular Weight Distribution at Low Conversion

The molecular weight of radical chain polymerization is complicated because several possible terminations–disproportions, coupling, and various transfer reactions are involved. Molecular weight changes with % conversion, due to the changes in the concentrations of monomer and catalyst, and the rate constant of propagation and termination.

At low conversion where all of the kinetic parameters $\left([M],[I],k_d,k_p,k_t\right)$ are approximately constant. Under these conditions, the polymer molecular weight does not change with conversion. The molecular weight distributions can be derived as did in the step polymerization and are shown in equations below, as the number-fraction, number-, and weight-fraction distributions, the number- and weight-average degrees of polymerization, and the breadth of the distribution, respectively.

$$\underline{N}_x = p^{x-1}(1-p)$$
$$N_x = N_0(1-p)^2 p^{x-1}$$
$$W_x = x(1-p)^2 p^{x-1}$$
$$\overline{X}_n = \frac{1}{(1-p)}$$
$$\overline{X}_w = \frac{(1+p)}{(1-p)}$$
$$\frac{\overline{X}_w}{\overline{X}_n} = 1+p$$

Disproportional or chain transfer or a combination of two: one polymer molecule is produced from each kinetic chain. When p is used as the probability to continue propagation, the molecular weight distribution can be shown in the following.

$$p = R_p / (R_p + R_t + R_{tx})$$

where R_p, R_t, and R_{tr} are the rates of propagation, termination by disproportionation, and chain transfer, respectively. The breadth of the size distribution $\overline{X}_w / \overline{X}_n$ is equal to 2 as p approaches unity. For termination by coupling (R_{tr} = 0 and R_t is the rate of coupling), a polymer arises from the combination of two kinetic chains, the size distribution is narrow.

Consider the probability $\underline{N}_{Y+Z}$ of formation of an x-sized polymer by the coupling of y- and z-sized propagating radicals. $\underline{N}_{Y+Z}$ is the product of the probabilities, $\underline{N}_Y$ and $\underline{N}_Z$, of forming the y- and z-sized propagating radicals.

According to equation. $\underline{N}_x = p^{x-1}(1-p)$, $\underline{N}_Y$ and $\underline{N}_Z$ are given in the following:

$$\underline{N}_x = (1-p)p^{y-1}$$
$$\underline{N}_Z = (1-p)p^{z-1}$$
$$\because y+z = x$$
$$\therefore \underline{N}_Z = (1-p)p^{x-y-1}$$
$$\text{Then } \underline{N}_{y+Z} = (1-p)^2 p^{x-2}$$

Equation $\underline{N}_{y+Z} = (1-p)^2 p^{x-2}$ gives the probability of forming an x-sized polymer molecule by only one of many equally probable pathways. There are (x – 1) possible pathways of producing an x-sized polymer molecule when x is an even number. Each of the pathways has the same probability. Therefore, the total probability $\underline{N}_x$ of forming an x-sized polymer molecule is given by,

$$\underline{N}_x = (x-1)(1-p)^2 p^{x-2}$$

Where $\underline{N}_x$ is mole or number-fraction distribution.

When x is an odd number there are x pathways,

$$\underline{N}_x = x(1-p)^2 p^{x-2}$$

For polymerization yielding in high polymer, the difference between x and (x + 1) is negligible and can be ignored. Then, we can have the following relationship:

$$W_x = \frac{1}{2}x(1-p)^3(x-1)p^{x-2}$$

$$\overline{X}_n = 2/(1-p)$$

$$\overline{X}_w = (2+p)/(1-p)$$

$$\frac{\overline{X}_w}{\overline{X}_n} = (2+p)/2$$

At high conversion, the size distributions become much broader than those for low concentration case. Kinetic chain length $\overline{v}$ is defined below:

$$\overline{v} = k_p[M]/2(fk_d k_t[I])^{1/2}$$

Then,

$$\overline{v} \propto [M]/[I]^{\frac{1}{2}}$$

The molecular weight is increased according to equation $\overline{v} = k_p[M]/2(fk_d k_t[I])^{1/2}$. Because [I] is decreased faster than [M], then the molecular weight of polymer is increased with % conversion. For polymer having practical usage, we want its PDI as narrow as possible. Sometimes, it is necessary to have multiple addition of initiator, monomer during the polymerization to minimize the molecular weight broadening. For auto-acceleration, the PDI is even larger with a value of 5–10. For chain transfer branch polymer, is also large with a PDI of 20–50, because chain transfer to polymer increases as the polymer size increases.

Coordination Polymerization

The reaction where vinylic monomers polymerize through coordination at the metallic center of some catalytic species is called coordination polymerization. Although the first catalytic system based on this kind of coordination chemistry was reported by Phillips Petroleum Co., most of the literature concerning coordination polymerization refers to the Ziegler–Natta catalysts because of their versatility in controlling chemical composition distribution (CCD) and of the wider variety of monomers they can polymerize.

The main early advances in the knowledge and understanding of coordination polymerization were established by studying the catalyst discovered by Ziegler in the early 1950s for ethylene polymerization, and utilized by Natta for polymerization of propylene and α-olefins. The most important contribution of Natta's works consisted in developing the Ziegler catalysts that could control polymer stereoregularity. Natta separated and characterized the three polypropylene stereoisomers—isotactic, syndiotactic, and atactic, thus opening the doors to a revolution on polyolefin applications that is still seen today.

Heterogeneous Ziegler–Natta Polymerization

Catalysts

The Ziegler–Natta initiators are the only initiators that polymerize α–olefins such as propene and 1-butene which can not be polymerized by either radical or ionic initiators. Thousands of different combinations of transition and Group I-III metal components, often together with other compounds such as electron donors, studied for use in alkene polymerizations.

Catalysts are prepared by mixing the compounds in a dry, inert solvent in the absence of oxygen usually at a low temperature. The mixture of aluminum compound with titanium compound is to form radical, the first of this kind catalyst being used as shown below.

$$AlR_3 + T_iCl_4 \rightarrow AlR_2Cl + TiRCl_3$$

$$AlR_2Cl + T_iCl_4 \rightarrow AlRCl_2 + TiRCl_3$$

$$AlR_3 + TiRCl_3 \rightarrow AlR_2Cl + TiR_2Cl_2$$

$$TiRCl_3 \rightarrow TiCl_3 + R.$$

$$TiR_2Cl_2 \rightarrow TiRCl_2 + R.$$

Further reduction may also occur:

$$TiRCl_2 \rightarrow TiCl_2 + R.$$

$$TiRCl_3 \rightarrow TiCl_2 + RCl$$

In addition, $TiCl_3$ may be formed by the equilibrium.

$$TiCl_4 \rightarrow TiCl_2 \rightleftharpoons 2TiCl_3$$

The stereo regularity of the polymers depends on the type of catalysts used as shown in table which indicates the isotacticity of polypropylene is varied by different catalyst. Syndiotactic polypropylene has been prepared under both heterogeneous and homogeneous conditions, but only under homogeneous condition a high degree of syndiotactic placement has been achieved.

Table: Variation of polypropylene isotaciticity with catalyst.

Catalyst[a]	Stereoregularity (%)
$AlEt_3 + TiCl_4$	35
$AlEt_3 + \beta\text{-}TiCl_3$	45
$AlEt_3 + \alpha\text{-}TiCl_3$	85

$AlEt_3 + ZrCl_4$	55
$AlEt_3 + VCl_3$	73
$AlEt_3 + TiCl_4$ + P,As or Sb compounds	35
$AlEt_2X + TiCl_3$	90–99
$AlEtX_2$ + γ-$TiCl_3$ + amine	99

- [a]Et = ethyl; X = halogen.

With the appropriate choice of catalyst, specific stereochemical arrangements of polymerization of diene monomer can be obtained as shown in tables. The 1,3-butadiene monomer will obtain cis-1, 4; trans-1, 4; isotactic 1, 2-, and syndiotactic 1, 2-polymers. The isoprene monomer will obtain cis- and trans-1, 4; 1, 2-; and 3, 4-polymers.

Table: Catalysts for the stereospecific polymerization of butadiene.

Catalyst[a]	Yield (%)	Polymer structure[b]
$R_3Al + VCl_4$	97–98	trans-1,4
$R_3Al + VCl_3$	99	trans-1,4
$R_3Al + VOCl_3$	97–98	trans-1,4
$R_3Al + TiI_4$	93–94	cis-1,4
$R_2AlCl + CoCl_2$	96–97	cis-1,4
$R_3Al + Ti(OC_4H_9)_4$	90–100	1,2
$Et_3Al + Cr(C_6H_5CN)_6$;Al/Cr = 2	~100	st-1,2
$Et_3Al + Cr(C_6H_5CN)_6$;Al/Cr = 10	~100	it-1,2

- [a]Et = ethyl.
- [b]st = syndiotactic, it = isotactic.

Table: Catalysts for the stereospecific polymerization of isoprene.

Catalyst[a]	Yield (%)	Polymer Structure
R_3Al + α-$TiCl_3$	91	trans-1,4
$Et_3Al + VCl_3$	99	trans-1,4
$Et_3Al + TiCl_4$; Al/Ti < 1	95	trans-1,4
$Et_3Al + TiCl_4$; Al/Ti > 1	96	cis-1,4
$Et_3Al + Ti(OR)_4$	95	3,4

- [a]Et = ethyl.

The cis polymer is usually formed at relative low temperature, for instance, polymerization of acetylene using an $Al(C_2H_5)_3 - Ti(OC_4H_9{-}n)_4$ catalyst yields cis polymer at temperatures below −78 °C, but trans polyacetylene is formed at 150 °C. The result may be due to thermal isomerization of the initially formed cis polymer.

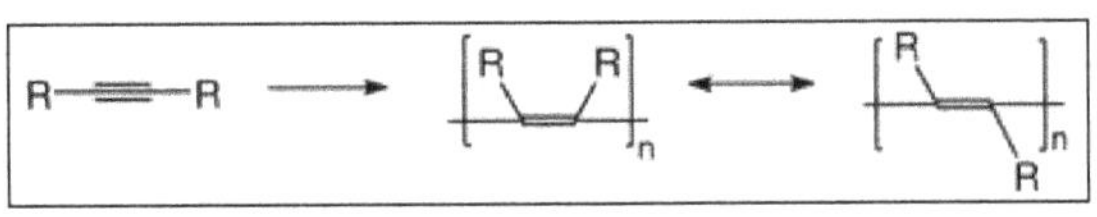

Reaction Mechanisms

The coordination polymerization of alkene can be preceded either by monometallic mechanism or bimetallic mechanism depending on the catalyst. Scheme shows the reaction mechanism of substituted alkene being polymerized by using titanium compound catalyst. The double bond of alkene will undergo cis addition and coordinate with the empty orbital of titanium compound to form four membered ring coordinate intermediate I. The stereo specificity of substituted alkene is preserved via intermediate I. Then the bond is formed on the Ti compound catalyst. If the catalyst is made from aluminum compound and titanium compound, the polymerization will proceed through bimetallic mechanism as shown in Scheme A bridge structure II is formed between two metal compounds first. Then the substituted alkene is coordinated with Ti compound to form six-membered ring coordination III. The stereo specificity of substituted alkene is preserved via intermediate III.

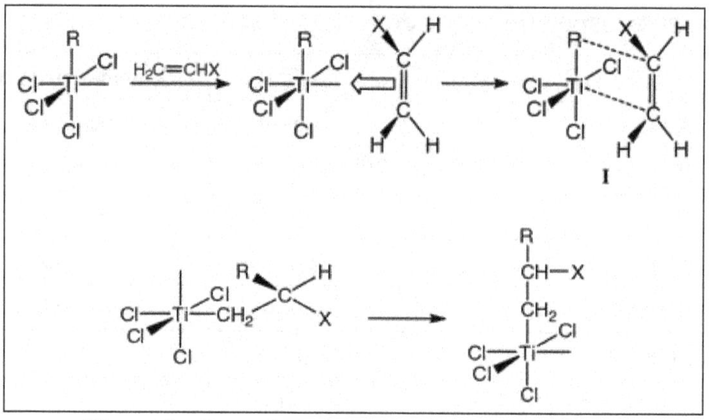

Scheme: Monometallic mechanism of Ziegler–Natta polymerization of substituted alkene.

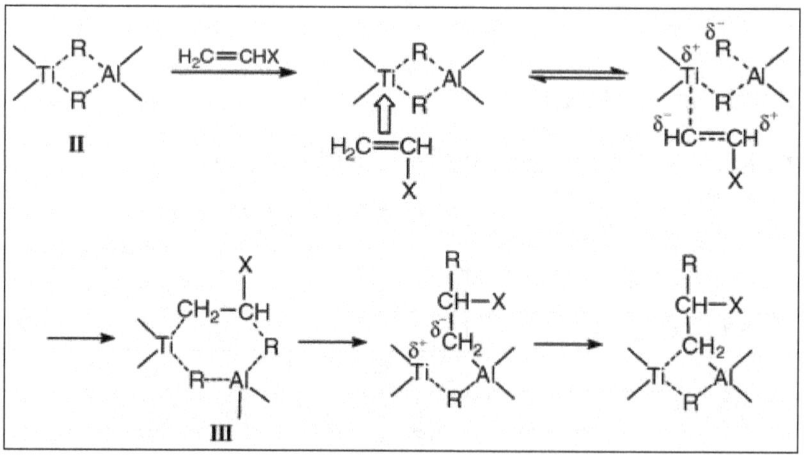

Scheme: Bimetallic mechanism of Ziegler–Natta polymerization of substituted alkene.

The polymer chain grows from the catalyst surface by successive insertion reactions of complexed monomer and the R group originally present in the organometallic cocatalyst ends up as the terminal group of the chain. Monomer activity decreases with increasing steric hindrance about the double bond as shown below:

$$CH_2 = CH_2 > CH_2 = CHCH_3 > CH_2 = CHCH_2CH_3 > CH_2 = CHCH_2CH(CH_3)_2 > CH_2 = CHCH(CH_3)_2 > CH_2 = CHCH(CH_2CH_3)_2 > CH_2 = CH - C(CH_3)_3$$

The reaction has the characteristic of living anionic polymerization. The reaction is usually terminated by active hydrogen as shown in the following Eqs. Hydrogen is the preferred transfer agent for controlling molecular weight due to low cost and clean reaction but the termination reaction is usually carried out by hydrogen containing compounds as shown in equation.

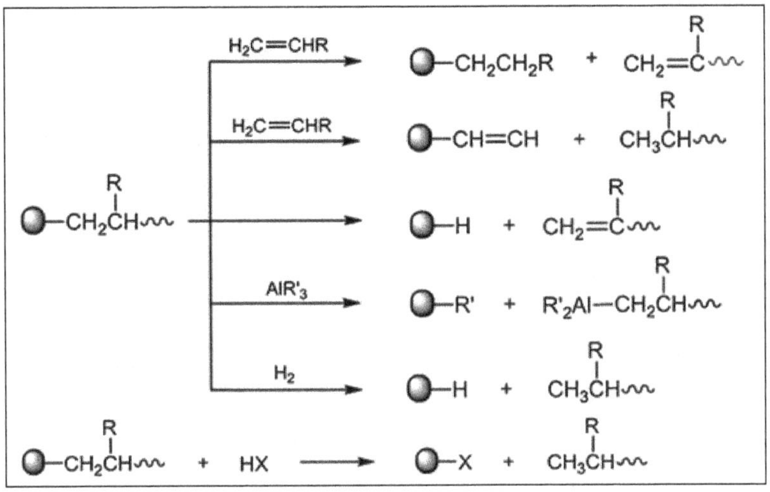

The relationship between the polymerization rate and time is shown in figure. The decaying rate type is most common. That is due to structural changes from the reducing the number or activity of active centers. It is also due to the encapsulation of active centers by polymer which prevents approach by monomer.

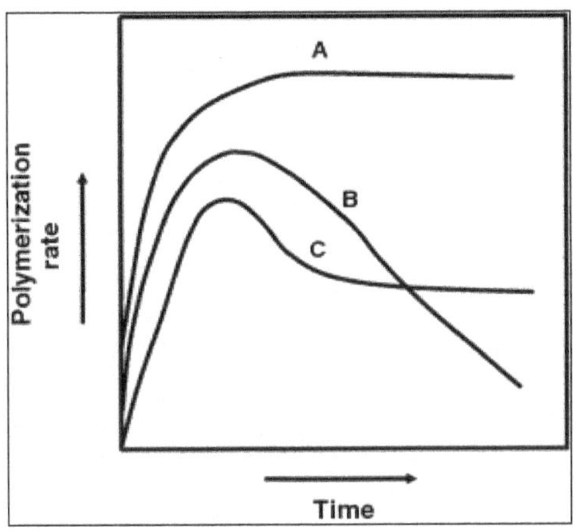

Three types of polymerization rate observed in Ziegler–Natta polymerization: a constant. b decaying. and c decaying to constant.

Molecular weight distributions are generally broad when insoluble catalysts are used and much narrower with soluble catalysts. The board distribution may arise from the decay of catalyst activity or from the presence of sites of variable activity.

There are two theories to explain the reaction mechanisms of coordination of dienes. One theory is based on whether the catalyst coordinates one or both double bonds of the diene. Coordination of one would thus lead to 1, 2-polymerization and coordination of both to 1, 4-polymerization. Another theory is based on the coordination of a π-allylic structure that directs the monomer approaching direction and determines the polymer structure. If the monomer approaches the CH_2-metal

(M) bond of the complex, 1, 4-polymerization forms. If it approaches the CH-metal bond, 1, 2-polymerization results. This mechanism provides no information on the geometric arrangement of the double bond or the tacticity at a stereogenic carbon.

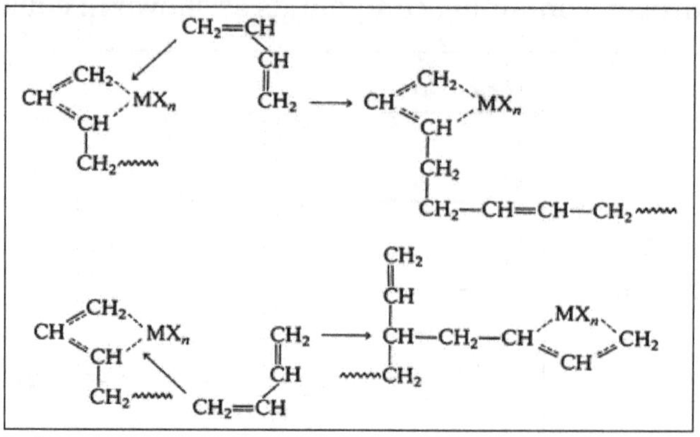

Scheme: Reaction mechanisms of coordination of dienes.

Homogeneous Ziegler–Natta Polymerization

The homogeneous Ziegler–Natta polymerization using metallocene catalysts such as bis(cyclopentadienyl)titanium dichloride 1, and dialkylaluminum chloride 2. Their structures are shown below.

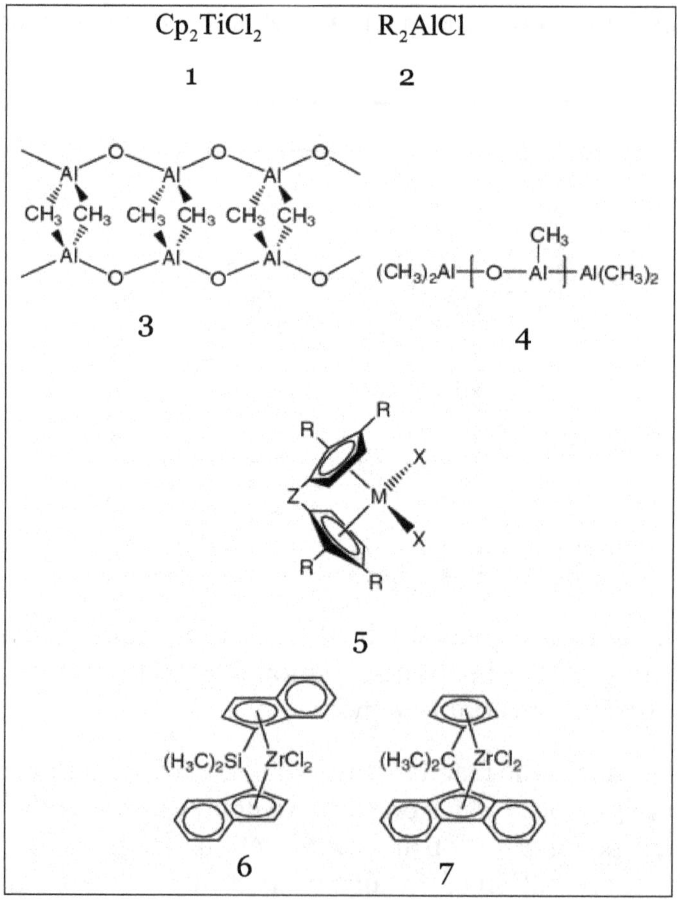

Compounds 1 and 2 exhibit low catalytic activity towards ethylene and are generally unreactive toward propylene. The addition of water increases the activity substantially. The increase is the result of a reaction between the water and the alkylaluminum cocatalyst to form complex alkylalumoxanes such as methyl alumoxanes (MAO). The MAO is used in conjunction with metallocene catalysts exhibit especially high activities. MAO formed by controlled hydrolysis of trimethylaluminum that has a complex oligomeric structure with molecular weights of 1000–1500, most likely consisting of methyl-bridged aluminum atoms alternating with oxygen as shown in structures 3 and 4. MAO is now used with a wide variety of metallocenes having the general structure 5. Examples of catalysts are 6 and 7 which form isotactic and syndiotactic polypropylene, respectively.

Metallocene has well-defined molecular structure and polymerization occurs at one position in the molecule, the transition metal atoms. Thus, the metallocene is also called *single-site catalyst* in contrast to the multi active site of heterogeneous catalyst. Scheme shows an example of the formation of active site in a zirconium catalyst, L_2ZrCl_2 (where L represents the π ligands) which involves initial complexation between MAO and the catalyst, is followed by Cl-CH_3 exchange to form $L_2Zr(CH_3)_2$. The methylated zirconocene reacts further with MAO to form the active species of 8.

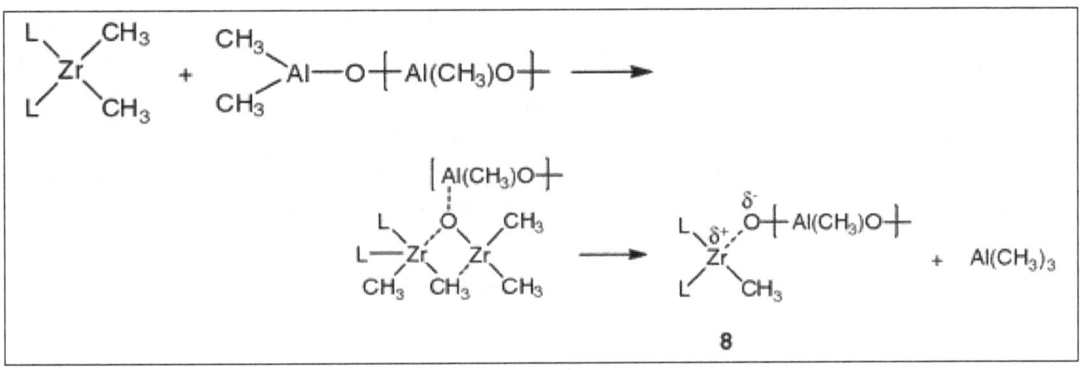

Scheme: Formation of the active site in a zirconocene catalyst.

Scheme shows the possible polymerization mechanisms of ethylene using zirconocene catalyst. The ethylene is cis added to the empty orbital of Zr site to form transition site, then is further on to precede polymerization.

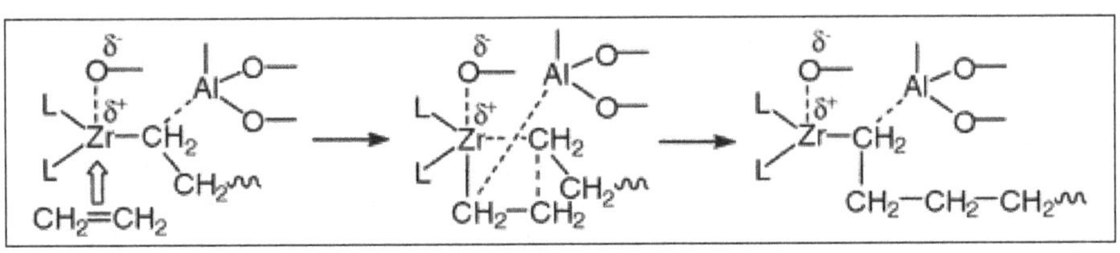

Scheme: Possible polymerization mechanism for ethylene using zirconocene catalyst.

The comparison between metallocene catalysts and heterogeneous catalysts for alkene polymerization is shown in Table. The metallocene out performs over the heterogeneous catalysts. Heterogeneous catalysts contain sites of variable activity that accounts for the large PDI. Metallocene catalysts differ from heterogeneous catalysts is that they are capable of polymerizing strained cyclo alkenes such as cyclobutene or norbornene 9 through the double bond to yield crystalline stereoregular polymers. Ring-opening polymerization of cycloalkene monomers is more common with heterogeneous Ziegler–Natta catalysts.

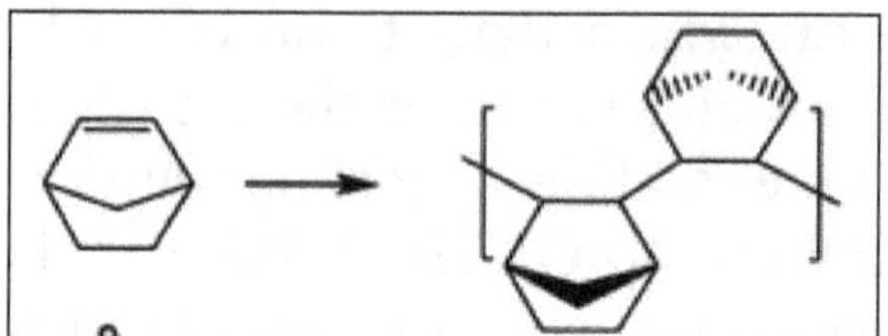

9

Table: Comparison between metallocene catalysts and heterogeneous catalysts for alkene polymerization.

Catalyst type	PDI	Catalyst activity	Strained cycloalkene
Metallocene	2–2.5	10–100	Yes
Heterogeneous	5–6	1	No

The molecular weight decreases with increasing polymerization temperature, with increasing catalyst concentration, with the addition of hydrogen to the monomer feed.

The reactivity of metallocene catalyst 10 depends on its chemical structure. The M of metallocene can be Zr, Ti or Hf. The X is Cl or alkyl; Z is an optional bridging group, $C(CH_3)_2$, $Si(CH_3)_2$, or CH_2CH_2; R is H or alkyl. For the group 4B metals, the order of activity is Zr > Ti > Hf. Alkyl groups on the cyclopentadiene rings increase catalyst activity if they are not too bulky. Large, bulky alkyl groups and electron-withdrawing groups decrease the activity. Increasing the size of the groups attached to the atom bridging, the cyclopentadiene rings (C or Si) reduces the activity. MAO affords much higher catalyst activities than ethyl or higher alkyl alumoxane cocatalyst.

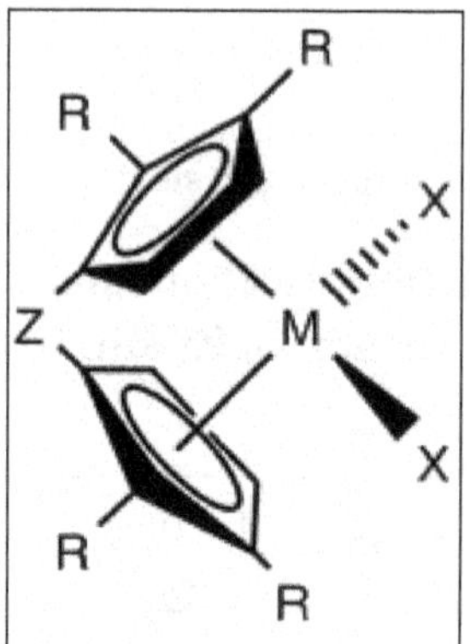

10

Achiral catalysts of $CpZrCl_2$, bridged or unbridged yield atactic polymer. Catalyst 11(chiral) yields isotactic; 12 (achiral) yields syndiotactic polymer. The much different size of the two π ligands of 12 is assumed to play a role in the formation of syndiotactic polymer. Substitution of a methyl group on the cyclopentadiene ring of 12 results in hemi isotactic polypropylene (mixture of isotactic and atactic). Catalyst 13 can rotate between chiral and achiral geometries, with the chiral form being responsible for the isotactic blocks and the achiral form for the atactic blocks. The isotactic blocks form crystalline microdomains in the polymer matrix, the stereoblock copolymers exhibit elastomeric properties. The degree of isotacticity is sensitive to propylene pressure and reaction temperature.

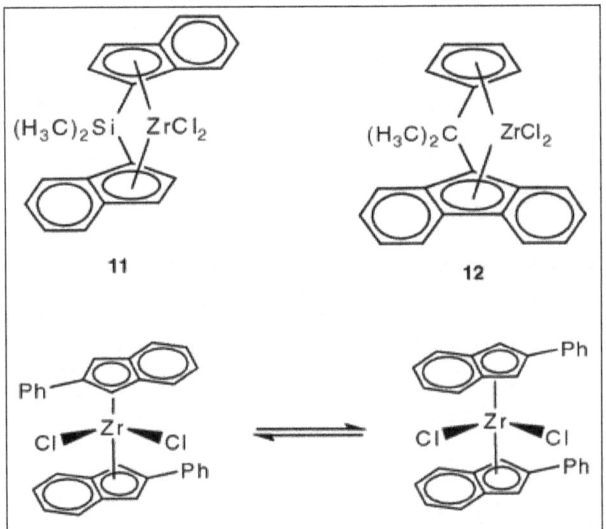

11 12 13 14

What are the reaction mechanisms for the stereospecificity of metallocene catalyzed polymerization? Most models take into account steric interactions involving the π ligands attached to the metal atom and the last inserted monomer unit of the polymer chain, as the incoming monomer inserts itself into the constrained active site. Scheme describes the isotactic placement for 11. Optically active isotactic polymer will form from pure enantiomer of 11. The larger fluorenyl group restricts the approach of the propylene molecule such that one side of the coordination site is enantioselective to one face of the monomer, and the other side of the coordination site shows the opposite enantioselectivity. Thus as the coordination site moves from side to side, syndiotactic polymer is formed. A methyl group on the cyclopentadiene ring presumably interferes with the selectivity on one side only. Since only one side of the coordination site is then stereoselective, a hemi-isotactic polymer forms.

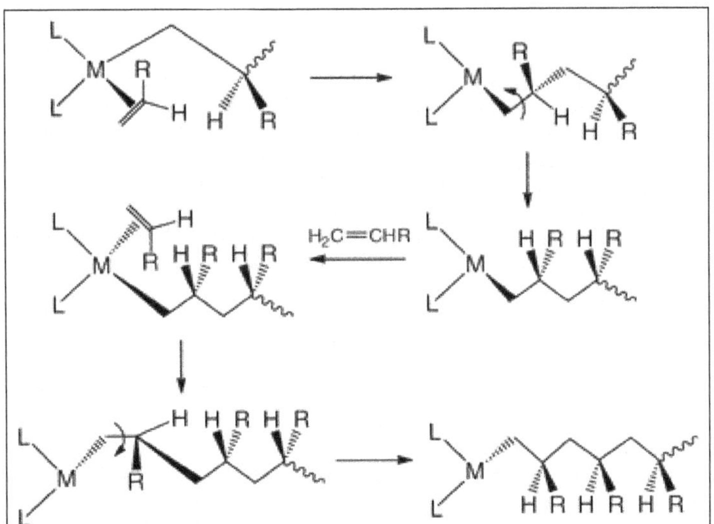

Scheme: A mechanism for isotactic placement with a metallocene catalyst.

Dow Chemical Company is using metallocene catalysis to produce its new Questra syndiotactic polystyrene (SPS). Dow has transformed polystyrene—an amorphous, low melting point, relatively brittle plastic—into a tough crystalline material with a melting point of 270 °C (conventional is 250 °C) for electronics, automotive under-the-hood, engineered films, and medical applications. Therefore polystyrene can be either low cost common plastic or high performance engineering plastics.

Ziegler–Natta Copolymerization

Random copolymers of ethylene and 1-alkenes are obtainable with Ziegler–Natta catalyst. The most important ones are ethylene with 1-alkene (LLDPE) and ethylene with ethylene propylene monomer (EPM) or ethylene-propylene-diene monomer (EPDM). Table summarizes the reactivity ratio of Ziegler–Natta copolymerization. The reactivity ratio is the rate constant ratio of monomer toward itself and toward other monomer and expressed as:

$$r_1 = k_{11} / k_{12} \text{ and } r_2 = k_{22} / k_{21}.$$

Table: Representative reactivity ratios of Ziegler–Natta copolymerization.

Monomer 1	Monomer 2	Catalyst[a]	r_1	r_2
Heterogeneous				
Ethylene	Propylene	$TiCl_3/AlR_3$	15.72	0.110
Ethylene	Propylene	VCl_3/AlR_3	5.61	0.145
Ethylene	1-Butene	VCl_3/AlR_3	26.90	0.043
Propylene	1-Butene	VCl_3/AlR_3	4.04	0.252
Homogeneous				
Ethylene	Propylene	Cp_2ZrMe_2	31	0.005
Ethylene	Propylene	$[Z(Ind)_2]ZrCl_2$	6.6	0.060
Ethylene	1-Butene	Cp_2ZrMe_2	55	0.017
Ethylene	1-Hexene	Cp_2ZrMe_2	69	0.020

- [a]R = C_6H_{13}; Cp = cyclopentadiene; Me = methyl; Z = CH_2CH_2; Ind = indene.

Ethylene is more reactive than higher alkenes with both heterogeneous and homogeneous catalysts. In most instances, $r_1 r_2$ is close to unity. With heterogeneous catalysts, a wide range of compositions is generally obtained, possible because different active sites may give rise to different reactivity ratios, or because of decay of activity. A more homogeneous polymer composition is obtained with soluble catalysts, particularly if monomer composition is carefully controlled to remain relatively constant during polymerization.

There is serious question as to whether the compositions also include significant amounts of homopolymer. The problem arises with uncertainties over the life-times of propagating chains, which may become detached from catalyst by chain transfer, thereby leaving the catalyst still active to initiate homopolymerization of a second monomer. Thus Ziegler–Natta method is inferior to anionic polymerization for synthesizing carefully tailored block copolymers.

Kodak's polyallomers and Uniroyla's TPR are block copolymer of ethylene and propylene. They are high impact plastics exhibiting crystallinity characteristics of both isotactic polypropylene and linear polyethylene. They contain homopolymers in addition to block copolymers. The conventional random copolymers of ethylene and propylene are elastomeric nature.

Metathesis Polymerization

Alkenes undergo a double bond redistribution reaction called olefin metathesis in the presence of certain complex coordination catalysts. Bond redistribution involves cleavage of the double bonds (trans-alkylidenation). Scission of the single bonds adjacent to the double bonds (trans-alkylation) has been ruled out as a possibility from deuterated alkene experiments. Tungsten compounds are particularly effective for olefin metathesis. The catalysts are formed by the reaction of tungsten hexachloride (WCl_6) with alcohol and ethylaluminum dichloride ($C_2H_5AlCl_2$). Other transition metals, notably Ru, Mo, Rh, and Ti are also effective.

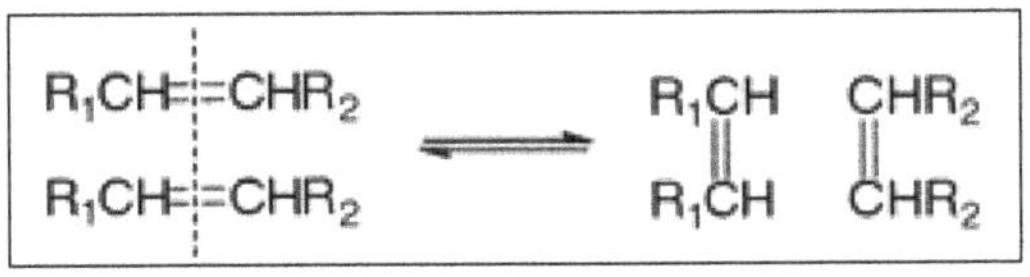

The reaction mechanisms of metathesis polymerization are shown in equation.

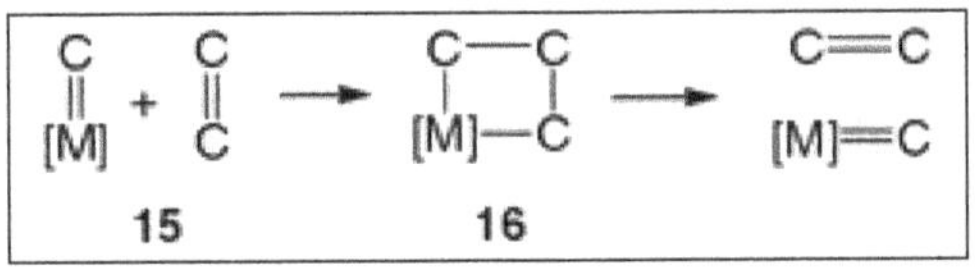

Equations show two examples of metathesis polymerization. A double bond containing polymer can be obtained through metathesis polymerization. The method developed in 1970, now it is recognized as a particularly convenient way to make polymers containing precisely spaced carbon–carbon double bonds in the polymer backbone.

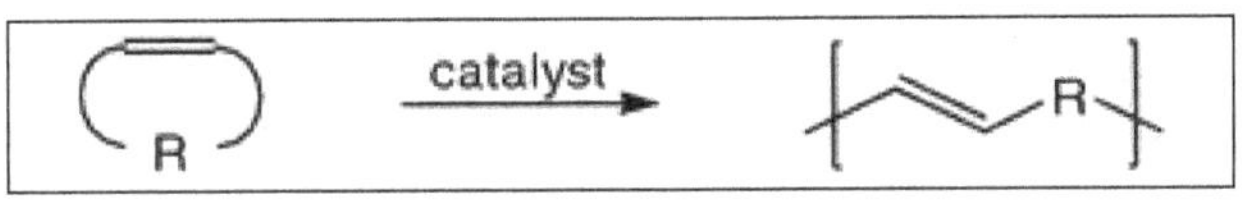

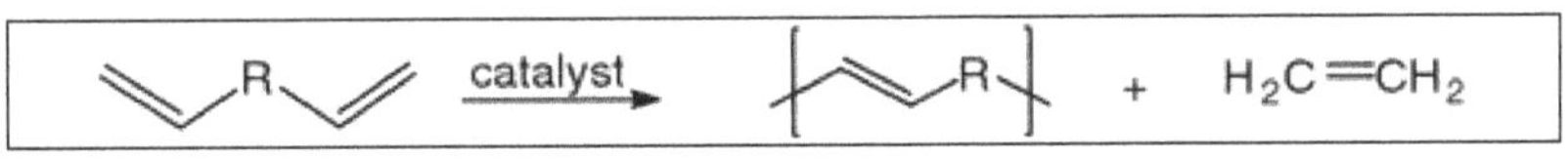

The Grignard metathesis reaction has been used to synthesize conducting polymer extensively such as poly(3-hexyl thiophene) as shown in figure.

References

- Polymerization, science: britannica.com, Retrieved 23 June, 2019
- Polycondensation: adhesiveandglue.com, Retrieved 3 August, 2019
- Materiale-didattico, allegati, webdocenti-be: docenti.unina.it, Retrieved 19 May, 2019
- Free-Radical-Polymerization, Polymerization, Reactions-of-Alkenes, Purdue-Chem-26100%3A-Organic-Chemistry-I-(Wenthold), Purdue: chem.libretexts.org, Retrieved 28 April, 2019

5

Stereochemistry of Polymers

Stereochemistry refers to the discipline of chemistry that studies the relative spatial arrangement of atoms which form the structure of molecules. This chapter closely examines the key concepts of stereochemistry of polymers to provide an extensive understanding of the subject.

Stereochemistry (also known as 3D chemistry) concerns the three dimensional spatial arrangement of atoms in a molecular structure and the resulting effects of these different arrangements on the reactivity and properties of the molecule.

When a polymer has stereochemical isomerism within the chain, its properties often depend on the stereochemical structure. Thus, the analysis of the Stereo-Chemistry of polymers is important and NMR spectroscopy has been the most valuable tool for this purpose. It is a general rule that for a polymer to crystallize, it must have highly regular polymer chains. Highly irregular polymers are almost inevitably amorphous. Polymer chains can have isomeric forms that decrease the regularity of the chains.

There are three important forms of isomerism in polymers.

- Structural Isomerism.
- Sequence Isomerism (Head-to-Tail or Head-to-Head).
- Stereoisomerism (Tactility).

Stereochemical formulae for polymer chains are shown as Fischer projections rotated through 90° i.e. displayed horizontally rather than vertically, or as hypothetical extended zigzag chains the latter occasionally give a clearer indication of the three-dimensional arrangement.

Structural Isomerism of Polymers

In the saturated hydrocarbons, whose structural formulae are shown in figure, it is not possible to form distinct isomers with just three or less carbon atoms linked together. There is only one way in which one carbon and four hydrogen atoms can be linked together, the single compound being methane, CH_4. A similar situation holds for ethane, C_2H_6 and propane, C_3H_8. But with butane, two possible structures can be formed: *n*-butane, which has a linear structure, and *iso*-butane, where the central carbon atom is linked to three adjacent carbons rather than one. Their physical properties are slightly different, for example their boiling points differ by 10 °C, but otherwise they are very similar compounds. Drawing the possible structures (or using structural models) of the isomers gives 3 isomers for pentane, 5 for hexane, 9 for heptane, and so on. As the number of possible structures increases, their properties diverge: the boiling points of the heptanes range over 20 °C and the melting points by no less than 110 °C. The chemical properties differ too: witness the different combustion behaviour of *n*- and *iso*-octane in petrol engines. The straight chain isomer causes 'pinking' during combustion, while the branched chain burns smoothly in the engine chambers (hence the familiar 'octane rating' for petrols).

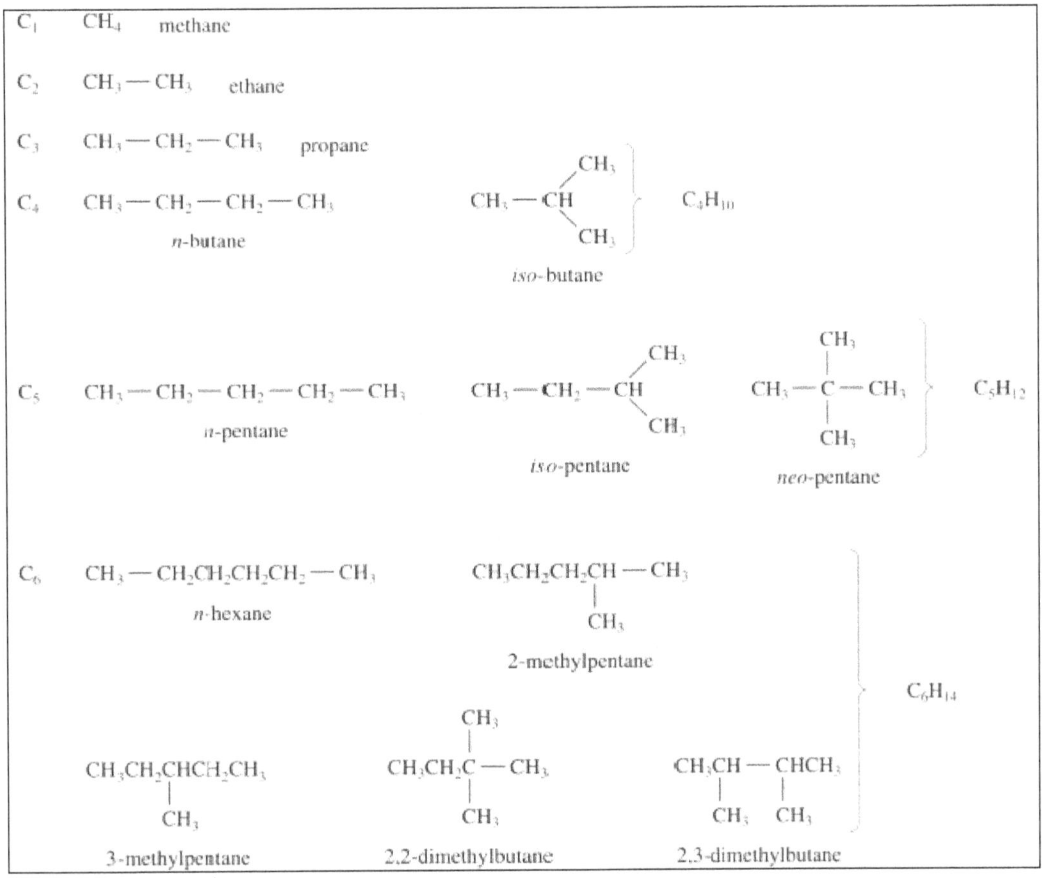

Configurational isomers of saturated hydrocarbons (alkanes): The number of isomers increases rapidly with the number of carbon atoms in the structure. There are no isomers for the first three members of the series, two for butane (C_4), three for pentane (C_5) and five for hexane (C_6).

The increase in isomeric structures is so rapid that at C_{30}, there are no less than 4 111 846 763 theoretically possible compounds. So with even the lowest molecular mass polyethylene, there is an almost infinite number of isomers. Fortunately, the situation is simplified enormously by the way polymerization occurs and in fact there are relatively few chemically distinct polyethylenes. the concept of a distinct molecular formula is redundant with most commercial polymers since chain lengths are very variable even within a single sample, but the idea of branching is important for polyethylene in particular.

Geometrical Isomerism

A second type of isomerism occurs with diene monomers, and is present in both NR and butadiene rubbers (BR). It occurs because the single double bond in the final polymer can exist in two ways: a cis form and a transform. The repeat unit shown in table for NR does less than justice to the two-dimensional structure of this material. In this planar formula where the bonds are shown in their correct orientation to one another, the base polymer in natural rubber can be seen to be m-polyisoprene, where the pendant methyl group appears on the same side as the lone hydrogen atom. The two parts of the chain in which this single repeat unit sits, lie on the opposite side of the double bond.

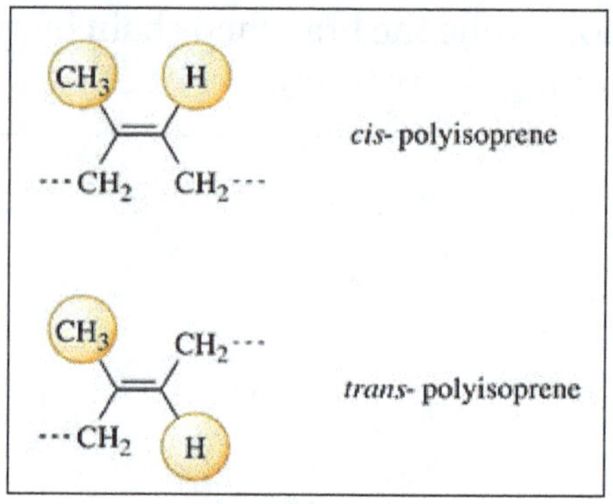

Geometrical isomerism in polyisoprene. Cis-polyisoprene in the main polymer occurs in natural rubber, trans-polyisoprene in gutta percha.

But there is an alternative structure with exactly the same formula: the polymer is gutta percha, Here, the two parts of the chain are on opposite sides of the double bond, giving the overall chain a zig-zag appearance. It is also a naturally occurring polymer, but has quite different properties to natural rubber. It is a highly crystalline and rigid material without the long-range elasticity characteristic of an elastomer. It was used formerly as an electrical insulator, but has now been largely superseded by synthetic plastics.

Sequence Isomerism of Polymers

In chain polymerization monomers with pendant groups can attach in two ways.

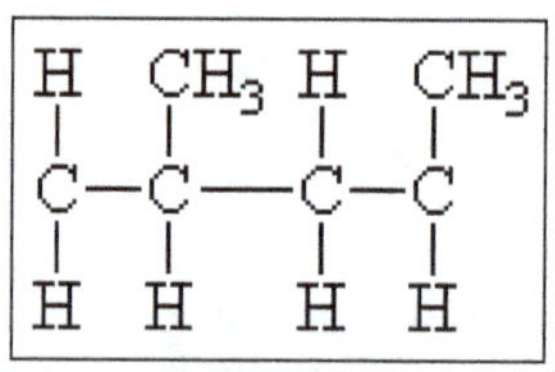

Head-to-tail.

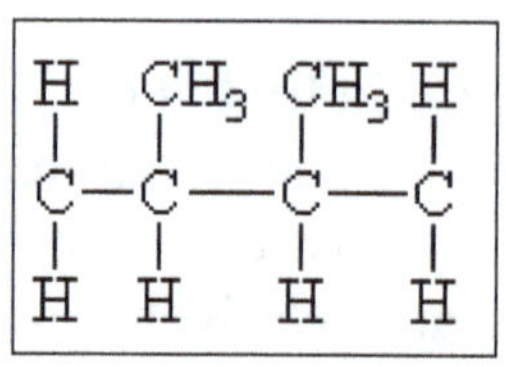

Head-to-head.

The usual arrangement is head-to-tail with the pendant groups on every other carbon atom in the chain.

Stereoisomerism of Polymers

A final type of isomeric variation occurs as a result of the three-dimensional structure of some polymers. It is possible because a four-valent atom like carbon can exist in two different forms when the subsidiary groups or atoms attached to the carbon are all different. The carbon atom is then known as an asymmetric carbon atom. A very simple example of the phenomenon is the structure of a small molecule, lactic acid. As figure shows, it can exist in two forms which are mirror images of one another. One of the two possible compounds, laevo- (standing for left-handed), or l-lactic acid occurs in muscle after vigorous, anaerobic exercise and causes muscle cramp. It is a good example of stereoisomerism in a small molecule, a feature it also shares with a large number of biological molecules, as well as some polymers.

The carbon atom in a vinyl polymer to which is attached the pendant side group (i.e. every alternate carbon atom in the main chain) is another example of an asymmetric carbon atom. It gives rise to tacticity. When the zig-zag chain is written with the plane of the chain in the plane of the paper, there are three ways in which the position of the pendant side group can exist. The methyl groups in polypropylene, for example, can occur all on one side (isotactic), on alternate sides (syndiotactic) or placed at random (atactic). These possibilities are shown in figure. The properties of each type of polymer are quite different to one another, primarily because isotactic and syndiotactic PP have ordered chains and so can crystallise, but atactic chains are quite irregular and cannot crystallise. Isotactic PP is the common form of the commercial material, although atactic PP is used as a binder for paper for example. Syndiotactic PP has recently become available commercially being made using a new family of catalysts, known as metallocenes.

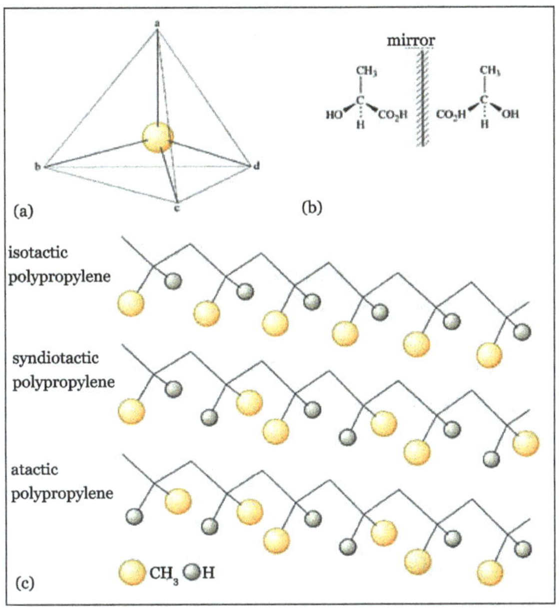

(a) Tetrahedral configuration of single carbon atom; (b) left and right-handed forms of lactic acid; (c) tacticity in polypropylene.

Structure of Polymers

Polymers are composed of non-metallic elements, found at the upper right corner of the periodic table. Carbon is the most common element in polymers. The chemical bonds in polymers are also different than those found in metals and ceramics.

Covalent Bonds

Non-metallic elements have a high number of valence electrons (four or more) and prefer to gain electrons, not lose them, in chemical reactions. They often form anions. In a compound of only nonmetals there are no elements willing to become cations, so ionic bonds are not possible. Instead, two nonmetallic atoms can share valence electrons with each other. This type of electron sharing, called covalent bonding, keeps the shared electrons close to both atomic nuclei. One pair of shared electrons makes one covalent bond. A molecule is a group of atoms held together by covalent bonds.

This type of bonding contrasts with metallic bonding, in which valence electrons are not associated with a particular nucleus, and move easily throughout a sample.

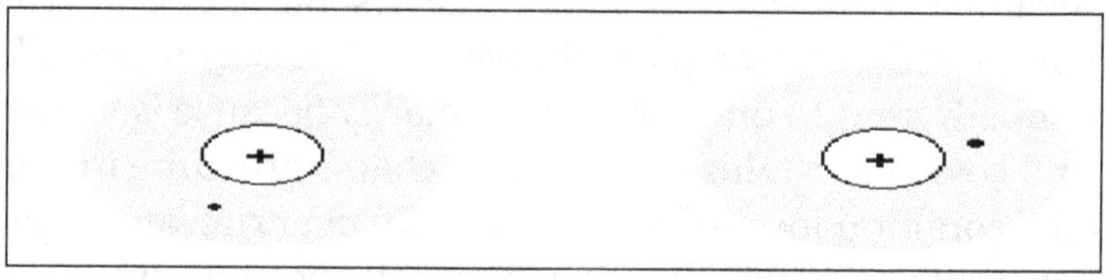

Each hydrogen atom has one proton in its nucleus, and one electron.
The electron moves randomly in a spherical space around the nucleus.

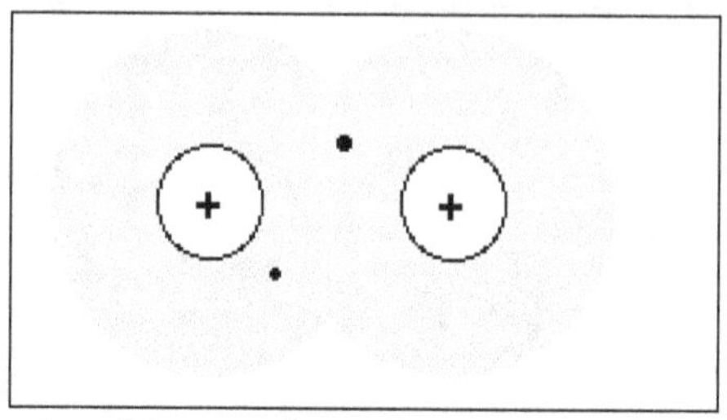

When a covalent bond forms, two hydrogen nuclei move close together.
Although the electrons still move around the molecule, usually they are near both nuclei.

To determine how many covalent bonds can be formed between atoms, first the number of valence electrons must be counted. This can be determined by using a periodic table. The group number matches the number of valence electrons.

Example:

- Carbon is element 6. It is found in Group IV so has four valence electrons.
- Oxygen is element 8. It is found in Group VI so has six valence electrons.
- The molecule carbon dioxide has the chemical formula CO_2.
- 4 electrons from C + 2(6) electrons from O = 16 valence electrons.

Lewis Structures

Rather than writing a sentence for the number of valence electrons on an atom, it can be more useful to draw a picture containing this information. A Lewis structure for an atom starts with a chemical symbol, with a dot added for each valence electron.

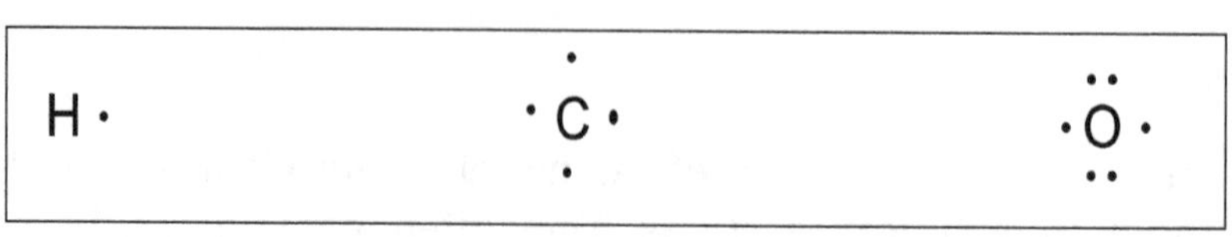

Since the highest possible number of valence electrons is eight (for noble gases) the dots representing valence electrons are traditionally arranged on four sides of the symbol, with at most two electrons on each side.

Experiments have shown that most nonmetallic nuclei are satisfied when they are near eight valence electrons. That is known as the octet rule. Carbon has four valence electrons, and needs to find four more to share. Oxygen has six valence electrons, so it only needs two more. Hydrogen is an exception to the octet rule; its nearest noble gas, helium, has only two electrons. Hydrogen nuclei form molecules with two nearby electrons, a duet rule.

To show a covalent bond, two chemical symbols are put near each other with two dots, representing a pair of electrons, between them. For example, a water molecule has one oxygen atom covalently bound to two hydrogen atoms.

water

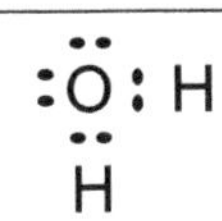

Nuclei do not have to share all of their valence electrons. Note that two pairs of the oxygen's valence electrons are not shared with any other atoms. Those electrons are called lone pairs or nonbonded pairs and can influence the chemical properties of a molecule.

Because dots can be difficult to see, it is common to draw a line segment for each bond (two electrons). The Lewis structure of a water molecule then looks like:

water

Hydrogen peroxide is a different compond of hydrogen and oxygen, with chemical formula H_2O_2. Since hydrogen atoms only need a duet of electrons they are found at the outside of the molecule; the two oxygen atoms need to be in the center where they can form more bonds. In this textbook that structural information will be given by underlining the central atom(s) in the chemical formula: H_2O_2.

hydrogen peroxide

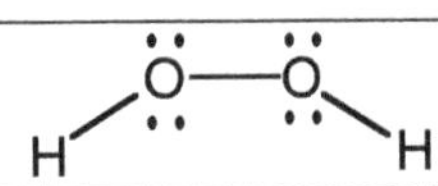

Sometimes nuclei will need to share more than one pair of electrons to achieve an octet with the available valence electrons. One shared pair of electrons is a single bond. Two shared pairs (four electrons) makes a double bond, and three shared pairs (six electrons) makes a triple bond. The more electrons that are shared, the stronger the bond will be. The neighboring elements carbon, nitrogen and oxygen commonly use double and triple bonds. For example, both nitrogen and oxygen are usually found as diatomic gases, N_2 and O_2. The nitrogen molecule has 2 x 5 = 10 valence electrons, and oxygen molecule has 2 x 6 = 12. Nitrogen needs a triple bond to achieve octets for each atom, but a double bond is sufficient for the oxygen molecule.

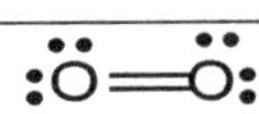

The type of covalent bond affects the shape of a molecule. The nuclei move closer together if they share more electrons. This means that a triple bond is shorter than a double bond, which is shorter than a single bond. The bond angles are also very specific in a covalently bond molecule. The shared electrons want to be near the two positively charged nuclei, but try to stay away from negatively charged lone pairs.

This structure is quite different than that created by metallic bonds, which do not have a particular orientation. Covalent molecules can flex a bit under stress but prefer to "bounce back" to their original positions.

Although a Lewis structure is a good way to show covalent bonds between atoms, it is not as effective at showing a molecule's three dimensional shape. Chemists use model kits and chemical graphics programs to visualize the positions of atoms in molecules.

Functional Groups in Polymers

Carbon is the most important element in polymers. Because it starts with only four valence electrons, and wants to share four more, carbon forms a wide variety of covalent bonds. Most importantly, carbon forms strong bonds with itself. Long, strong chains or nets made of thousands of carbon atoms form the backbone of a polymer.

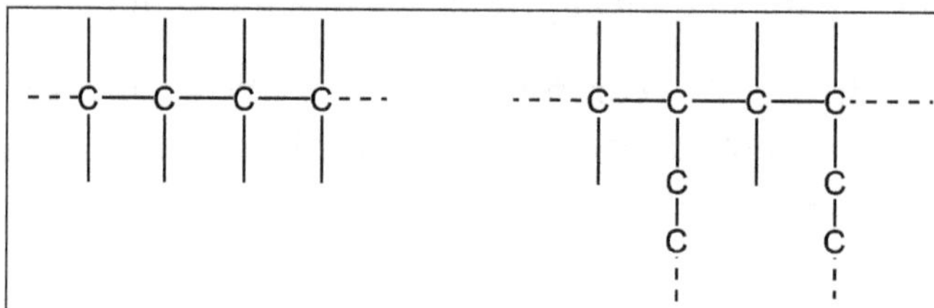

carbon backbones

Polyethylene is the simplest polymer. In addition to the carbon backbone, only hydrogen atoms are used to achieve four covalent bonds per carbon atom.

polyethylene

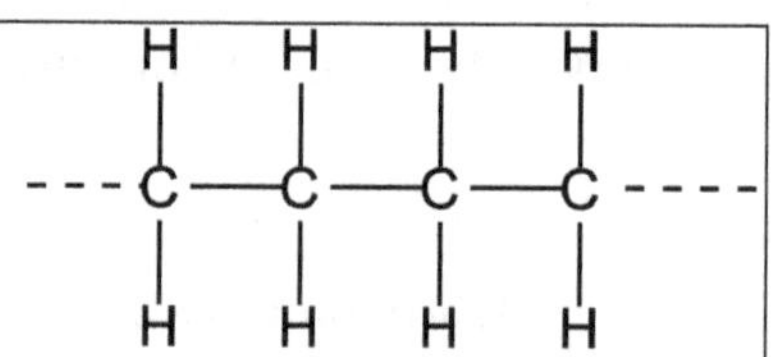

Although silicon is in the same group as carbon, it does not form strong bonds with itself. Silicones, long chains of alternating silicon and oxygen atoms, can be synthesized.

silicone

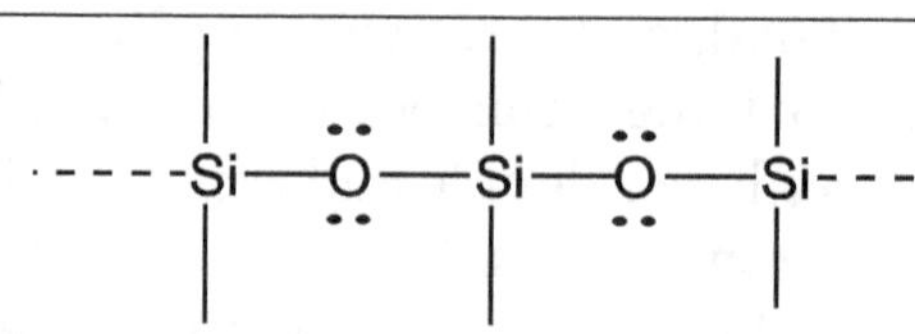

Many different nonmetal atoms could be covalently attached to a polymer backbone. Groups of atoms that contribute something besides C-C and C-H bonds are called functional groups. They affect the chemical and physical properties of a polymer. Examples of Functional Groups:

acid group (-COOH)

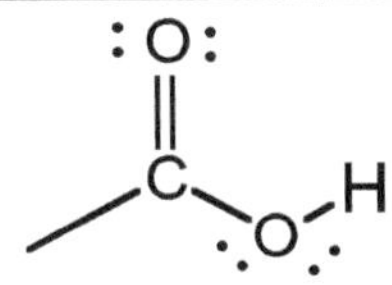

alcohol group (-OH)

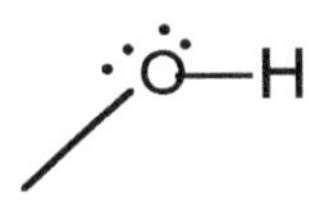

amino group ($-NH_2$)

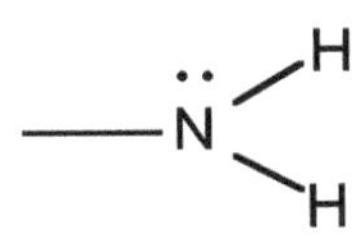

chloride (-Cl) fluoride (-F)

styrene ($-C_6H_6$)

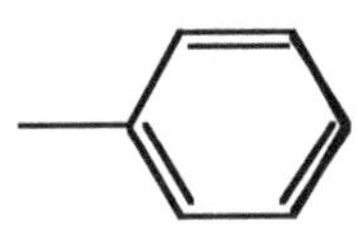

The primary experimental method used to identify functional groups in polymers is Infrared Spectroscopy (IR).

Skeleton Structures

Simplified or "skeleton" structures can be used to emphasize the functional groups. Carbon-carbon bonds of the framework are represented by line segments. Each vertex is the location of a carbon atom. Most hydrogen atoms and all lone pairs are omitted. This type of diagram deemphasizes the hydrocarbon skeleton; since it is so strongly bonded as to be unreactive, it does not affect the chemical properties of the polymer.

Polyethylene is the simplest polymer. Since it has no functional groups, the skeleton structure of a polyethylene fragment looks like it does not have any atoms. (Remember that a real polyethylene molecule is more often 100 or 1000 atoms long.)

polyethylene

It is possible to figure out the missing information. There should be a carbon atom at the end of each line segment; six are needed, connected by five single bonds.

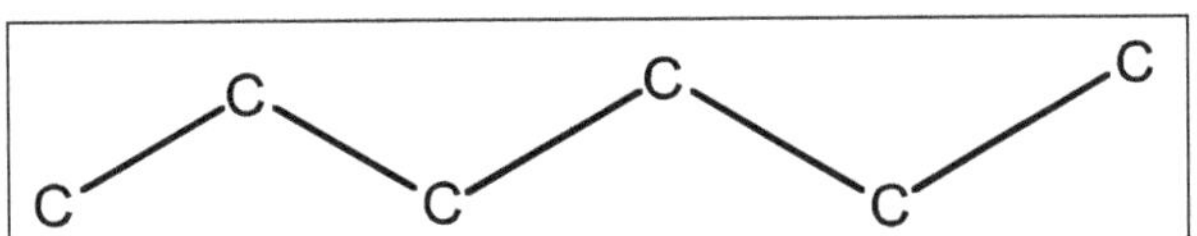

Since each carbon atom must have four bonds in a molecule, there must be missing bonds to hydrogen atoms. For the carbon atoms on the ends of the molecule, adding three C-H bonds to each will achieve octets. Two C-H bonds should be added to each of the inner carbons.

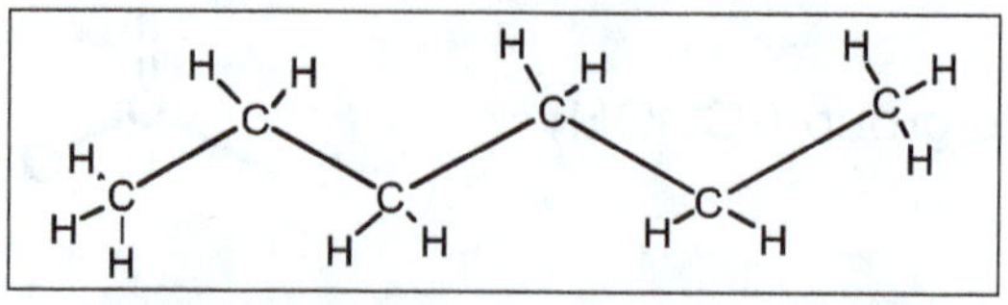

Complete Lewis structure for polyethylene fragment.

Three dimensional model of polyethylene.

When functional groups are added to a simplified backbone it is easy to notice the change in structure. Polyfluoroethylene, often sold as Teflon, is similar in structure to polyethylene except that all the hydrogen is replaced with fluorine. It is a very slippery polymer.

polyfluoroethylene

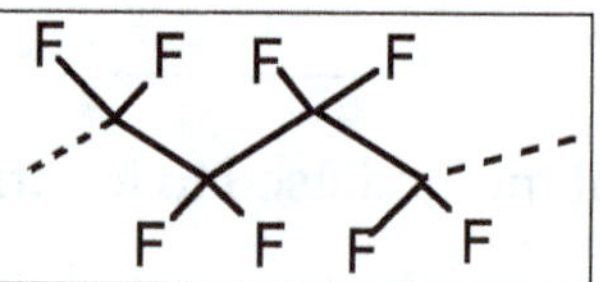

Amino acids, the monomers that build proteins, contain amino groups and acid groups, separated by one carbon. In the model nitrogen atoms are blue and oxygen atoms are red. The carbon between the amino group and the acid group always has one hydrogen on it (pointing up in the model) but the fourth group is variable. The symbol R is used when the exact identity of the group is not important. More than twenty different groups, as simple as a single hydrogen atom, are found on amino acids in nature.

amino acid

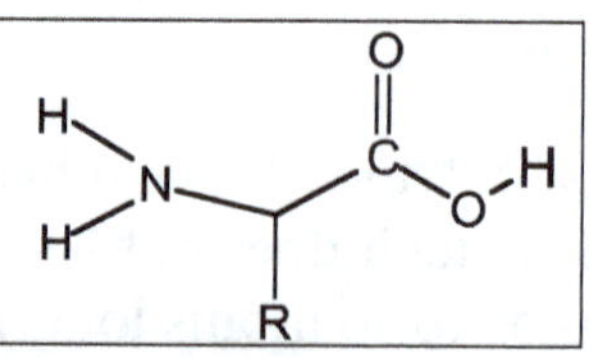

The photo shows a model of alanine, which has a $-CH_3$ group.

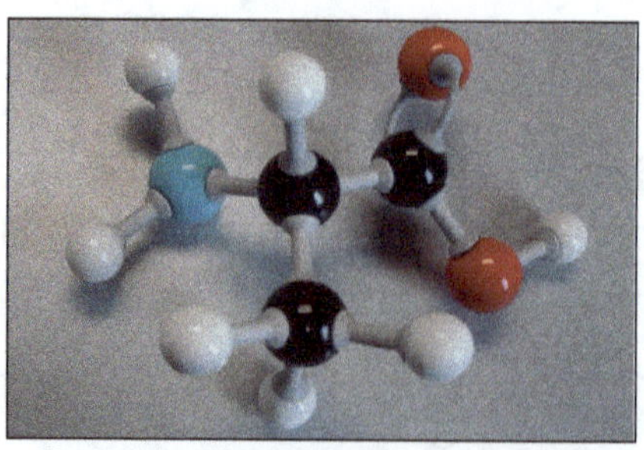

Three dimensional model of amino acid alanine.

Polysaccharides are sugar polymers. Cellulose (found in wood, cabbages, cotton, and linen) is composed of long chains of sugar rings. They are covered with alcohol groups.

cellulose

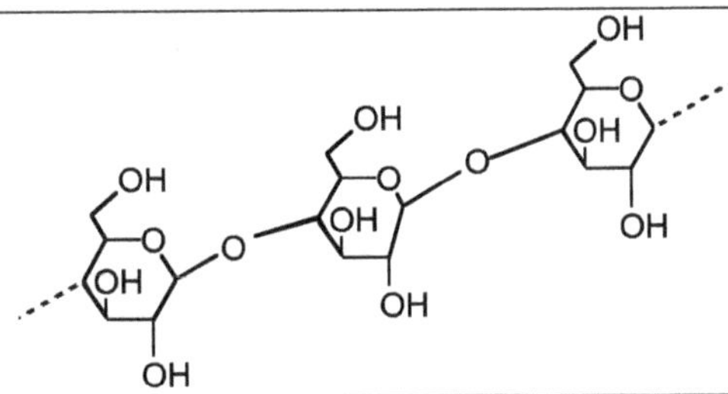

Starch contains the same functional groups, but the sugar rings are connected at different angles. The structural change makes it possible for the human digestive system to digest the polymer into sugar. Starch polymers are often branched.

starch

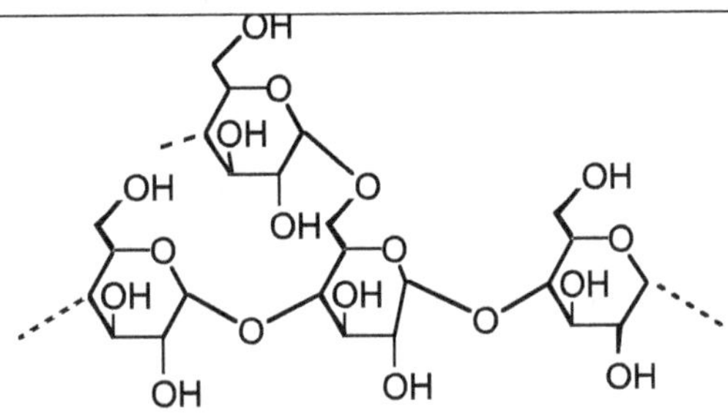

Intermolecular Forces

A molecule is a group of atoms connected by covalent bonds. Chemical reactions are required to form or break covalent bonds. Weaker attractions often form between molecules, encouraging them to stick together in groups. The weaker attractions are called secondary bonds or intermolecular forces. These can be overcome by adding heat or dissolving in a liquid. The functional groups on a polymer determine the type(s) and strength of its secondary bonds.

Polar Interactions

The valence electrons moving around a molecule may not be symmetrically distributed. The nonmetallic elements closest to the right top corner of the periodic table - nitrogen, oxygen, fluorine and chlorine - tend to shift shared electrons away from carbon and hydrogen. When there is a functional group with one of those elements, it has a slight negative charge and the rest of the molecule (carbon and hydrogen) is slightly positive. The molecule is polarized (or polar, for short). Its positive sections are attracted to negative sections of neighboring polymers.

poly(ethylene terephthalate)
"PET"

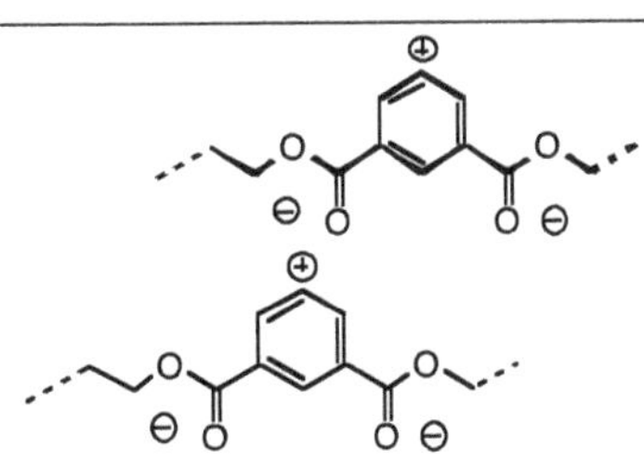

Poly(ethylene terephthalate) or PET, a polymer used to make bottles for carbonated beverages, has oxygen-containing functional groups that make it polar. Protein and cellulose chains are also polar.

Polyfluoroethylene is nonpolar (not polar) because it is completely covered with fluorine atoms; there is no exposed positive section to interact with a neighboring molecule's negative section.

Positive and negative charges can be localized on a covalent molecule since they have no path for conduction of electrons. The carbon atoms in the backbone always follow the octet rule with four covalent bonds, so can't pass extra electrons along the chain. If polymer fibers are rubbed together they can build up a static electricity charge.

Hydrogen Bonds

Molecules with either -N-H or -O-H groups will form strong secondary bonds. This phenomenon is responsible for the relatively high boiling point of water, and for the fact that its solid form (ice) is less dense than its liquid form. Polymers with hydrogen-bonding groups will soak up water.

Fabric softeners are added to laundry to change the properties of cotton and linen fabrics. The fabric softener molecules have one end that binds to OH groups on cellulose. The other end of the fabric softener is a long, nonpolar chain. This exposed end feels smooth and slippery. Softened fabrics are less able to build up a static charge. However softened fabrics will not absorb as much water. This is an issue for the performance of cotton towels.

Nonpolar Interactions

As valence electrons move around the nuclei in a nonpolar polymer, like polyethylene or polyfluoroethylene, they can become temporarily imbalanced. For a brief moment of time one part of a molecule would be negative, another part positive; it is temporarily polar. These occasional imbalances are enough to allow nonpolar molecules to attract each other, but the interaction is much weaker than that observed for polar or hydrogen bonding polymers.

Representations of Polymer Structures

Polymer structures are normally drawn as follows by showing only one repeating unit. Each representation below is equivalent to the corresponding structure.

polystyrene

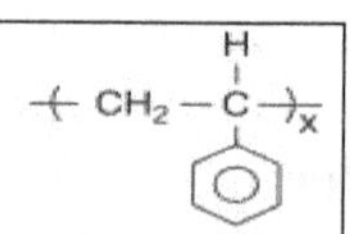

poly(ethylene terephthalate)

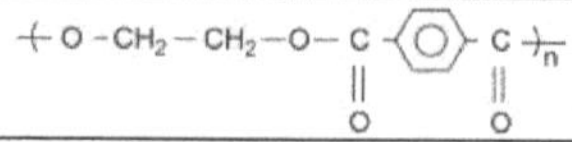

poly(hexamethylene adipamide)

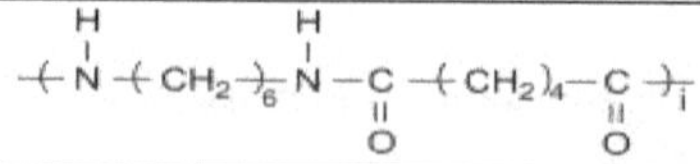

poly(vinyl alcohol)

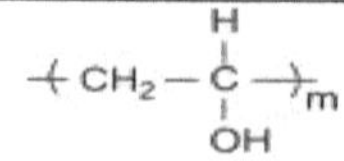

The subscripts x, n, i, m, and so on represent the number of repeating units in the polymer molecule. This number is often not known definitely in commercial synthetic polymer samples.

This representation of polymer structures implies that the whole molecule is made up of a sequence of such repeating units by linking the left-hand atom shown to the right-hand atom, and so on. Thus, the following structures are all equivalent to 1-5.

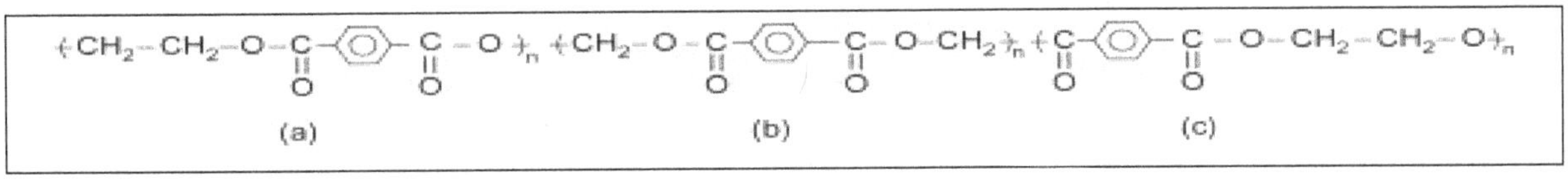

Formula (b) would not normally be written because the nominal break is in the middle of a —OCH_2CH_2O— unit, which is the residue of one of the precursors ($HOCH_2CH_2OH$) actually used in syntheses of this polymer by the reaction:

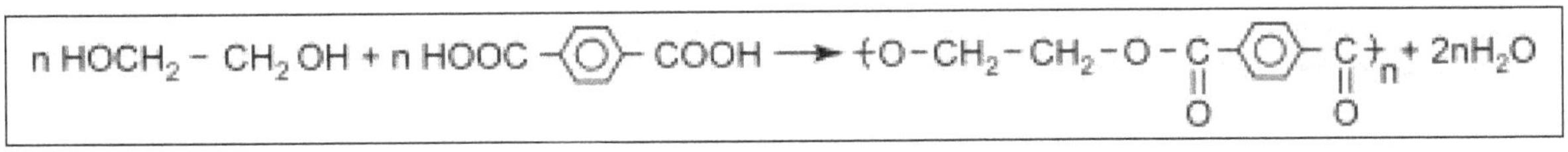

Configuration and Conformation

Steric Arrangement in Cis and Trans Configuration

Polymers with double bonds in the repeat unit give rise to different geometric isomers. For example, 1,3-butadiene can be polymerized to give poly(1,2-butadiene) or either of two geometric isomers of poly(1,4-butadiene). These two isomers are called cis and trans poly(1,4-butadiene). In the case of cis-polybutadiene, the first and the forth carbon lie on the same side of the central double bond, and in the case of trans, on opposite sides of the central double bond.

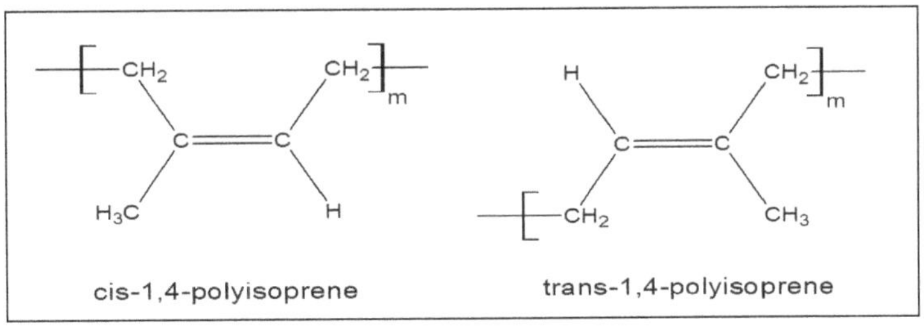

Double bonds in the cis form reduce the energy barrier for the rotation of adjacent C-C bonds compared to those in trans from, which, in turn, reduces the glass transition temperature (T_g). The affect of cis-trans isomerie on the T_g is usually not very strong, that is, the T_g shift is only about 20 K for polybutadiene. If other substituents (-Cl, -CH_3 etc.) are present, the energy barrier for the rotation of adjacent bonds will increase, and hence the T_g will increase as well. Bulky substituents will also reduce the energy barrier for rotation of the trans form relative to the cis form. For example, the observed T_gs of cis and trans polyisoprene are nearly the same, whereas the T_g of the cis isomer of poly(chloro-butadiene) is about 20 K larger than the T_g of the trans isomer.

Table: Glass transition temperature of polybutadienes.

Polymer	$T_{g,cis}$ (K)	$T_{g,trans}$ (K)
Polybutadiene	170 (165)	187 (190)
Polyisoprene	204	207
Poly(chlorobutadiene)	252	234

Double bonds in the polymer backbone have a singnificant impact on the polymer properties. The double bonds in the polymer backbone prevent rotation that occurs in carbon-carbon single bonds. It is believed that the inability to rotate freely hinders the allignment of the polymer chains. Thus, unsaturation in the polymer chain reduces the crystallinity and increases the elasticity and flexibility of polymers.

Tacticity in Polymers

Tacticity should not be confused with the conformational states of the polymer chains in space. The conformation refers to different arrangements of atoms and/or substituents of the polymer chain brought about by rotations about single bonds. Examples of different polymer conformations include the fully extended planar zig-zag, helical, folded chains, and random coils, etc. By contrast, the tactic configuration of the molecular chains refers to the organization of the atoms along the chain and configurational tactic isomerism involves the different structural arrangements of the atoms and substituents in a polymer chain, which can be interconverted only by the breakage and reformation of primary chemical bonds. There are three tactic forms in polymers: atactic, isotactic, and syndiotactic. Isotactic and syndiotactic polymers are both stereoregular and thus are crystallizable. Atactic polymers, on the other hand, are usually completely amorphous, unless the side group is so small or highly polar as to permit crystallinity, e.g. poly(vinyl fluoride) (PVF) or polyacrylonitrile (PAN). Like some organic compounds or naturally occurring polymers such as poly(L,D-lactic acid)s or poly(amino acid)s that differ in chirality, vinyl polymers differ in tacticity, which may be viewed as a pseudochirality form.

Tacticity in polymers is part of stereochemistry of organic materials. Differences in either tacticity or other types of stereoregularity in natural polymers are also common. Such concepts are widely used by nature to produce a variety of compounds of different properties or functions from same chemical species. Examples of different geometric.

Tacticity in Vinyl Polymers

Stereoregularity are seen in natural rubbers versus gutta percha, cellulose versus starch, etc, and differences of molecular stereoregularity in these natural polymers can result in significantly different mechanical and physical properties. In general, unlike difference in chirality that causes only optical properties but little differences in other physical properties in molecules or polymers, by contrast, differences in tactic or geometric stereoregularity in synthetic polymers or natural polymers result in significantly different properties. Tacticity in polymers causes phenomenal differences in microstructures, morphology, crystalline polymorphism, and physical/mechanical properties. Generally, tacticity in polymers is not 100%. A percentage of a specific tacticity can usually be characterized by using high-resolution solid-state nuclear magnetic resonance. Isotacticity in three consecutive monomer sequence is the placement called a meso-(same side) or m-placement; the syndiotactic structure corresponds to a racemic-(opposite) or r-placement. Each of these tactic combinations can be distinguished through NMR techniques.

Tacticity in Polymers

Tacticity in polymers is one type of stereoregular forms in molecular configurational structures. Polymers with specific tactic configurations can be viewed as pseudochiral polymers. The tactic polymers are, however, not really chiral and do not have chirality at all. Tacticity configuration is seen in vinyl polymers with mono- or disubstituted groups. The polymerization of monosubstituted ethylene, such as a vinyl compound, leads to polymers in which every other carbon atom is a potential chiral center. This center is marked with an asterisk, as follows:

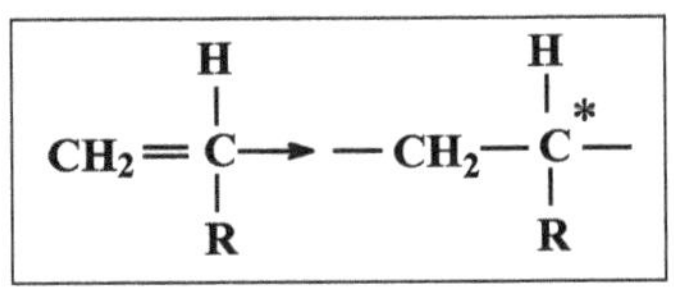

Such carbon atoms are referred to as pseudochiral centers in long-chain polymers because the polymers do not in fact exhibit optical activity. The reason for the lack of optical activity in tactic polymers can be seen through a closer examination of the substituents on such pseudochiral center:

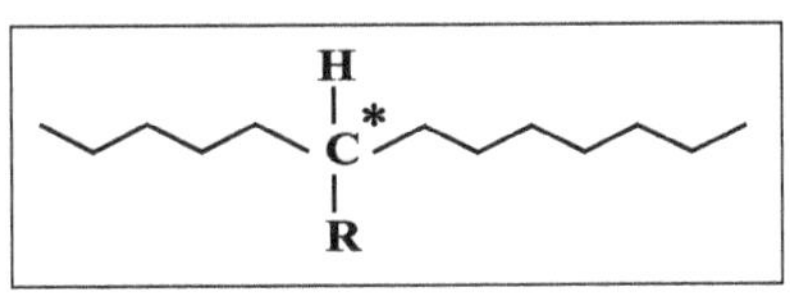

The two chain segments to left and right sides of the center carbon are in general of unequal segmental length, and thus they are structurally different "groups." However, in reality, only the first few atoms of the two chain segments attached to C∗ are responsible for the optical activity, not those farther away. These neighboring atoms to C∗ are seen to be the same, and hence the polymer is optically inactive (unable to cause optical rotation), even though the segments to the right and left-hand sides of C∗ are different.

Spatial Three-Dimensional View of Tacticity: Using Fischer– Hirshfelder or similar models, the actual differences between isotactic and syndiotactic vinyl-type polymers can be illustrated as where the dangling bonds on two ends refer to continuing main backbone.

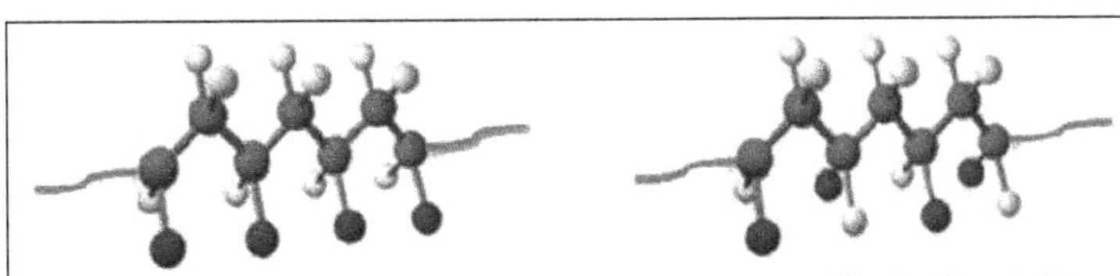

Isotactic configuration   Syndiotactic configuration

In the three-dimensional (3-D) schemes, the medium-size blue ball indicates a substituent group and the others are carbon and hydrogen atoms. For isotactic polymers, the substituent groups are all on the same side of carbon all-trans zigzag plane; for syndiotacticity, the substituent groups are alternatively on the opposite side of the carbon all-trans zigzag plane.

A two-dimensional (2-D) analogue can be made using Fisher projections. In these 2-D projections, the R groups are placed either up or down. All up- (or all down-) positions of R indicate the isotactic structure:

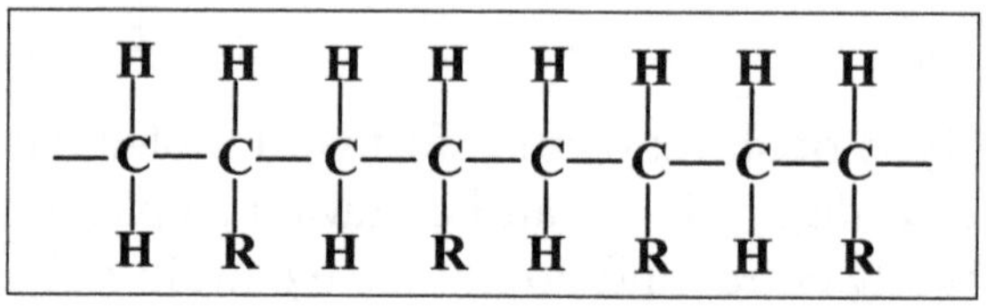

Alternating up and down positions indicate the syndiotactic configuration:

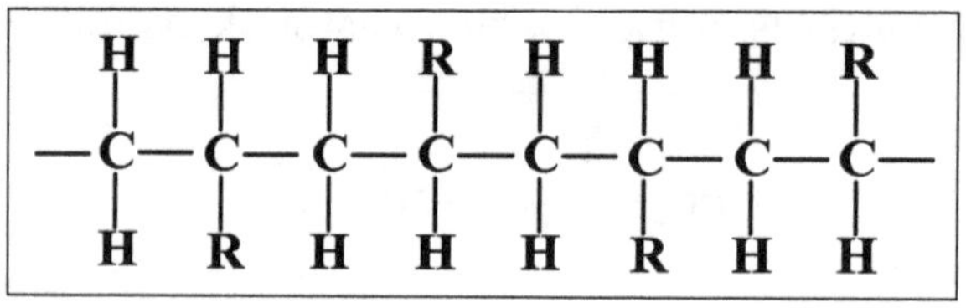

And random up and down positions indicate the atactic configuration:

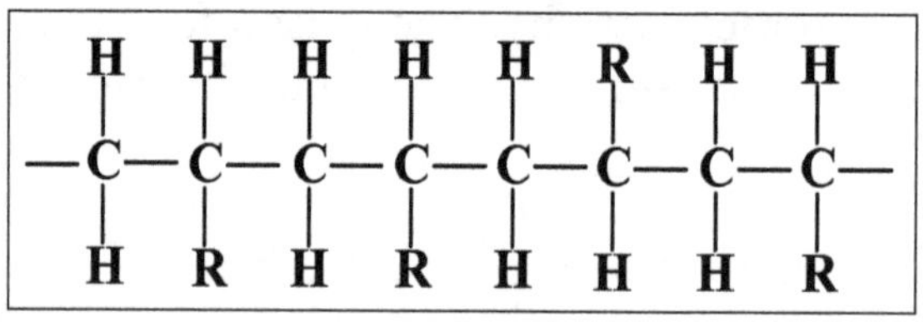

In specifying the tacticity of the polymer, the prefixes it- (or i-) and st- (or s-) are placed before the name or structure to indicate isotactic and syndiotactic structures, respectively. For example, it-polystyrene (or i-PS) means that the polystyrene is isotactic. The term i-PVC [poly(vinyl chloride)] means isotacticPVC, and so on. Such tactic polymers are known as stereoregular polymers. The absence of these tactic terms in polymer names denotes the corresponding atactic structures; for example, PS or poly(methyl methacrylate; PMMA) means atactic PS or atactic PMMA, respectively, and they are usually noncrystalline and amorphous owing to the lack of an stereoregularity order. Tacticity usually is seen in monosubstituted vinyl polymers, although in principle 1, 2-disubstitued polymers may also display four different forms of ditactic regularity. Ditactic 1, 2- disubstituted polymers, however, are not as common as monotactic polymers (tactic vinyl polymers with a single substituent). Tritacticity in polymers is also possible when the monomer units contain two tactic sites plus an additional double bond, for example, resulting in a total of three tactic sites.

Tacticity and Properties of Polymers

The structures as shown in the schemes for different tactic configurations may result in profoundly different physical and mechanical properties in polymers. The isotactic and syndiotactic structures are both crystallizable because of their regularity along the chain. However, their unit cells and melting temperatures are not the same. Atactic polymers, on the other hand, are usually completely amorphous, unless the side group is so small or the polymers are so polar as to permit some crystallinity, for example, in PVF. Tacticity in molecular structures influenced the glass transition and melting points. Syndiotactic polymers usually have a higher T_g than the isotactic forms of the same polymer. For example, isotactic PMMA (i-PMMA; weight-average molecular weight M_w = 300,000) has a T_g = 52 °C, but syndiotactic PMMA (s-PMMA; Mw = 50,000) has a much higher T_g = 124 °C. Polymer chains are obviously more flexible, thus have a lower T_g, when the pendant group is all aligned on the same side (isotacticity). Syndiotactity in polymers leads to higher T_g than isotacticity. Atactic PMMA (a-PMMA) has a T_g in between these two tactic forms, with T_g = ~100 °C.

Experimental Techniques and Apparatus for Tactic Polymers Characterization

Instrument for characterizing tactic polymers covers a variety of spectroscopic, thermal analysis, and microscopic techniques. Some of most commonly used are briefly listed here. Differential scanning calorimetry (DSC) is widely used for characterizing thermal properties, glass transition, crystallization kinetics, crystal polymorphism, and melting/crystallization transitions of tactic polymers. Wide-angle x-ray diffraction is widely used for determination of crystallinity and polymorphism, and crystal cell analysis on tactic polymers samples, with copper K_α radiation and a wavelength of 0.1524 nm. The scanning 2θ angle typically ranges from 5° to 40° with a step speed of 2° min^{-1}.

Microscopic characterization is utilized for analyzing the polymorphism, lamellar or spherulitic morphology of interior and surface of polymers. Polarizedlight optical microscopy (POM), usually equipped with a CCD (charge-coupled device) digital camera and a microscopic hot stage, is very useful for characterizing the phase domains or crystalline morphology, especially spherulites or dendrites, of tactic polymers. Growth kinetics and rates analyses of spherulites can also be measured by using automated software of image processing for POM. Transmission electron microscopy (TEM) or scanning electron microscopy (SEM) is used to characterize the crystalline lamellar morphology of tactic polymers. For TEM, microtoming of samples into ultrathin films using a diamond knife is required. Sometimes, samples may also be prepared as ultrathin films by solvent casting and are carefully picked up by using a copper grid. For SEM, the surfaces (deposited on glass slide or silicon) must be coated with vapor-deposited gold using vacuum sputtering prior to SEM characterization. Atomic-force microscopy (AFM) is a relatively newer microscopic characterization technique. An intermittent tapping mode is more common for obtaining the AFM phase and height images. For polymers as soft matters, the scan range of the AFM scanner is better to cover as much as 150 μm × 150 μm, with a narrower area of 5 μm × 5 μm for selective focusing into spots of interest. The modern-day AFM software is capable of tracking along a marked line for height profiling and a variety of useful analyses to compile images for surface crystal morphology or phase domains in tactic polymers.

Fourier-transform infrared spectroscopy (FT-IR) is used for investigating crystal forms, polymorphisms in tactic polymers. Spectra may be obtained at 4 cm^{-1} resolution, and averages of spectra are from at least 64 scans (for enhanced signals) in the standard wave number range 400–4000 cm^{-1}. Samples for IR measurements are cast as thin films with a uniform thickness directly on KBr pellets at ambient temperature, and IR measurements are performed on the samples cast on KBr pellets. For determining the tacticity and other configuration forms, solid-state ^{13}C nuclear magnetic resonance (^{13}C NMR) or solution-state ^{1}H NMR techniques (also Fourier transformed) are powerful tools. Tetramethylsilane is used as an internal standard for chemical shift references. Quantitative analysis based on the chemical shifts related to specific configurations in the tactic polymers can be performed to reveal tacticity forms and tactic percentages in tactic polymers.

Tactic Poly(methyl Methacrylate): a-PMMA, i-PMMA and s-PMMA

Vinyl polymers, when synthesized using suitable catalysts, can be produced to assume a fixed configuration, called tacticity, which is one type of stereoregularity. PMMA, poly(vinyl chloride), polypropylene (PP), poly(1-butene), PS, and so on are some common examples showing different tacticities, or low tacticity. Polymers of no tacticity (i.e, atactic polymers) are generally amorphous. Polymers of sufficient tacticity degrees are usually crystallizable, with the crystallinity depending on the degrees of tacticity.

In 1956, polymer of methyl methacrylate (MMA) was first synthesized using anionic polymerization by Szware and Rembaum. In 1958, stereoregular PMMA was prepared by Fox and co-workers. They reported that the stereoregular PMMA prepared with 9-fluorenyl lithium in 1,2-dimethoxyethane at −60 °C was used as an initiator. In 1983, another type of living polymerization for stereoregular PMMA, known as group transfer polymerization (GTP), was reported.

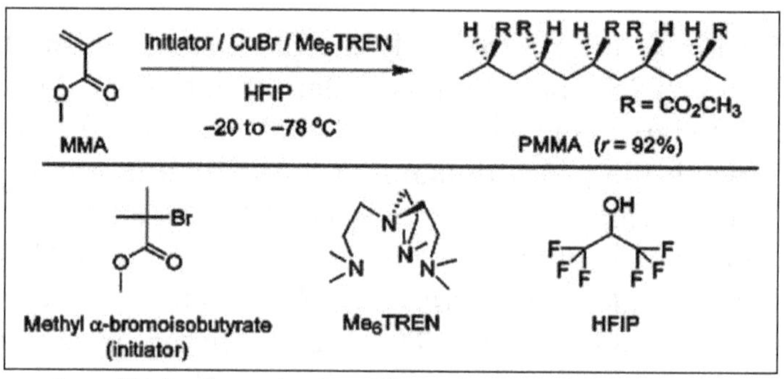

Atom-transfer radical polymerization design for controlling the M_w and tacticity of PMMA. (Postnote: r = 92% in figure may be rr = 92%.).

The narrow molecular weight distribution (MWD) of stereoregular PMMA can be produced by using the GTP method. More recently, the controlled/living radical polymerization was reported for controlling stereoregular polymers with targeted molecular weights and narrow MWD by Matyjaszewski and co-workers and Kakuchi and co-workers. For example, Kakuchi and co-workers reported that the simultaneous control of the molecular weight and stereoregular PMMA was successfully accomplished by using the atom transfer radical polymerization with the methyl αbromoisobutyrate/copper(I) bromide (CuBr/Me_6TREM) imitating system in HFIP at low temperature as shown in figure. However, the specific control of tacticity by strongly polar F-alcohol may be interesting and novel, but some investigators questioned that such a phenomenon might be due to low temperature used in synthesis.

Tacticity and Crystalline Properties of PMMA

Since the first report on preparation of stereoregular PMMA by Szware and Fox, three types of isomeric forms were found in PMMA, such as i-PMMA, s-PMMA, and a-PMMA. A wide variety of properties of PMMA is displayed depending on degree of tacticity, such as glass transition temperature (T_g), thermal degradation (T_d), specific heat (C_p) and entropy, and viscoelastic properties. i-PMMA shows a T_g around 50 °C and melting temperature (Tm) around 150 °C depending on degree of tacticity, whereas s-PMMA shows a higher T_g around 120 °C. Since a-PMMA is mainly composed of syndiotactic units, the T_g of a-PMMA at around 100◦C is much closer to the T_g of s-PMMA but far away from that of i-PMMA (T_g = 50 °C). Other properties, such as crystal structure, conformational energy, permeability properties, surface activity, and adsorption behavior, are also influenced by the type and degree of tacticity of PMMA. Several crystal structure models of i-PMMA have been proposed. Tadokoro and co-workers reported that i-PMMA structure was formed by adopting a 10_1 double-stranded helix as a molecular model and crystallized in an orthorhombic lattice cell with the parameters a = 41.96 Å , b = 24.34 Å , and c (fiber axis) = 10.50 A. By comparison, s-PMMA does not crystallize in melt crystallization at Å all, indicating poor crystallizability, but it may show a well-oriented crystalline structure by preparing in some specific organic solvents.

Effects of Tacticities on Phase Behavior of Blends

Over the past few decades, numerous reports on miscibility of polymers of different tacticities have been published. The miscibility of a polymer with another polymers of tactic forms has been reported to be different, such as blends of atactic PS with tactic poly(viny1 methyl ether; PVME). It has been found that isotactic PVME (i-PVME) was less miscible with atactic PS (a-PS) than syndiotactic PVME (s-PVME) with a-PS. Blends of atactic PVME with tactic PS (a-PS and s-PS) have also been investigated. A study by Woo and Mandal has concluded that the behavior of syndiotactic polystyrene (s-PS)/PVME blend agrees mostly with that of a-PS/PVME, indicating that tacticity or crystallinity in s-PS does not affect the phase behavior of s-PS/PVME in the amorphous domain. The phase behavior of a blend of PVME with s-PS tends to be quite similar to that of a blend of PVME with a-PS, indicating that tacticity in PS does not influence much the phase behavior in blends where the intermolecular interactions between two polymers are not based on configurational differences. The influence of tacticity on the phase behavior may be dependent on the interaction strength between two polymers, and effects of tacticity alone may be less influential when other factors are stronger.

Effects of Tacticity on the Phase Behavior in Blends

The miscibility of a-PMMA with other polymers is often similar to that of blends of s-PMMA with the same polymers; however, the T_g–composition relationships may differ slightly. This is quite understandable because a-PMMA is mainly composed of a significant percentage of syndiotactic units in the main chains. Some blends with a-PMMA or s-PMMA are miscible based on the observation of composition-dependent glass transition temperature (T_g) in blends, but blends with i-PMMA are immiscible, such as poly(Nvinyl pyrrolidone), poly(ethylene oxide) (PEO), and poly(styrene-coacrylonitrile) blends. Other factors, such as the mixing ratio, solvent effects, and molecular weight of the polymer constituents, also influence the miscibility of tactic PMMA blends.

Owing to increasing importance of biodegradable polymers, blends of biodegradable polymers, such as poly(L-lactide) (PLLA) or poly(hydroxylbutyrate) (PHB) with PMMA have been widely studied. Biodegradable polyesters (PLLA or PHB) are often modified with PMMA to improve their properties for service environments.

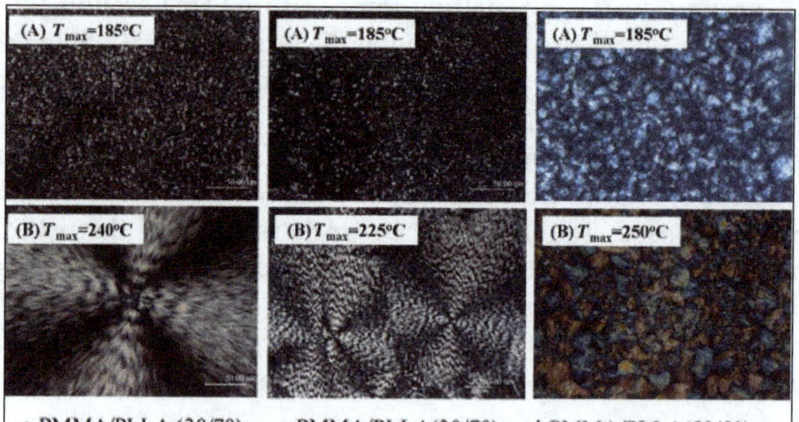

POM graphs of PLLA spherulites in blends of various tacticities
PMMA melt from the temperature below UCST (A) and above UCST (B). T_c = 130 °C.

However, the phase behavior and miscibility of biodegradable PLLA with tactic PMMAs can be ambiguous and remain to be quite controversial, owing to difficulty in interpreting the true phase behavior. A more valid interpretation on the phase behavior in PLLA/PMMA blends has been recently offered by Li and Woo, and they have claimed that the blend systems of sPMMA/PLLA and a-PMMA/PLLA are immiscible with an asymmetry-shaped upper critical solution temperature (UCST) behavior near 230–250 °C. On the other hand, an i-PMMA/PLLA blend remains immiscible up to thermal degradation without showing any transition to UCST upon heating. In addition to the phase behavior in blends, the morphology and nucleation density of PLLA crystals in tactic PMMA/PLLA blends are also affected by the critical temperature of melt treatment (Tmax) as shown in figure. This figure shows POM micrographs of the spherulites of PLLA blended with various tactic PMMAs melt-crystallized at 130 °C from the temperature below UCST (row A) and above UCST (row B). Figure shows small and irregular PLLA crystals with many occluded phases, which reflect a mechanism of heterogeneous nucleation. The large-sized PLLA spherulites are observed in figure, indicating that PLLA is crystallized from the single-phase liquid blend (above UCST).

Phase Behavior in Blends of a-, i- and s-PMMA

Generally polymers with same repeated units but different tacticities (i.e, stereoisomers) are not all miscible with each other. Examples are reported in blends of PMMA of different tacticities. White and Filisko have concluded that a-PMMA and i-PMMA could be one or two phases, as verified by using DSC, with the results depending on sample preparation methods. Krause and Roman earlier reported that a binary blend of i-PMMA and s-PMMA was miscible by using dilatometry analysis. However, Bauer and Bletso concluded differently that the i-PMMA/ s-PMMA blend system was immiscible, which was based on characterization results of dilatometry, mechanical damping, and torsional modulus. Schroeder and co-workers indicated that when i-PMMA and s-PMMA samples were mixed by using comparable molecular weights based on DSC measurements, miscible systems were formed in a whole range of compositions. These studies have shown that the phase behavior in blends of PMMAs remains yet unresolved and perhaps may be conflicting based on interpretations by different investigators.

Using optical and scanning electron microscopes, dynamic mechanical analysis, and DSC, effects of tacticity on the phase behavior in three pairs of binary blends comprising any two of i-, s-, and a-PMMAs have been thoroughly studied. Miscibility is confirmed in all three binary blends (i.e, a-PMMA/i-PMMA, a-PMMA/s-PMMA, and i-PMMA/s-PMMA); however, T_g versus composition relationships for the a-PMMA/i-PMMA blend differed significantly from the other two pairs. Tacticity in PMMA may lead to subtle differences in molecular interactions in the tactic polymers; and in turn such a difference in the interactions results in variation in the glass transition and phase behavior in these three blends. Chang and Woo have reported that miscibility is confirmed more unambiguously for all three pairs of binary blends (a-PMMA/i-PMMA, a-PMMA/s-PMMA, and i-PMMA/s-PMMA), which all exhibit a single, composition-dependent glass transition and a homogeneous phase morphology. Figure shows plots of T_g versus composition relationships that differ among the three pairs. By judging from the T_g behavior, the order of interaction in blends as follows: i-PMMA/s-PMMA > a-PMMA/s-PMMA ≫ a-PMMA/i-PMMA blends.

Complex Formation in Mixtures of PMMA of Opposite Tacticities

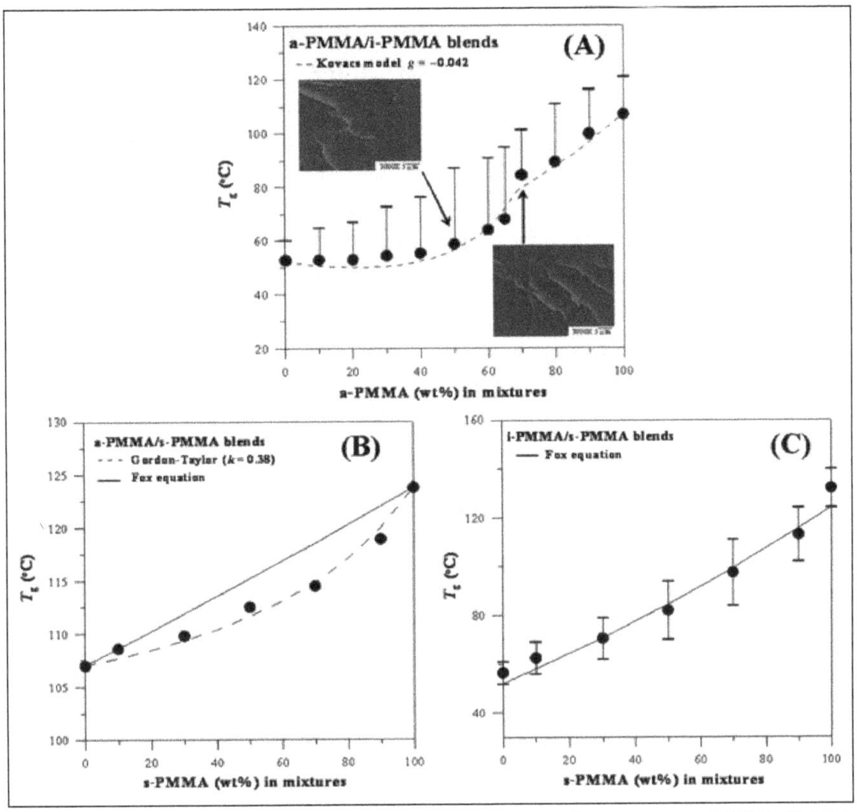

Plots of T_g versus composition for (A) a-PMMA/i-PMMA (B) a-PMMA/s-PMMA and (C) i-PMMA/s-PMMA blends.

Mixtures of polymers of same structures but opposite chirality are known to form complexes, such as complexes of PLLA and poly(D-lactic acid). Mixtures of polymers of opposite tacticities have also been reported to have capacity of forming complexes. Formation of stereocomplexes between i-PMMA and s-PMMA was first reported by Liquori and co-workers. Other factors, such as the polymer concentration, mixing ratio, molecular weight, and solvent effect of the stereoregular polymers, also affect formation of stereocomplexes between i-PMMA and s-PMMA. In addition, stereocomplexes composed of tactic PMMAs are known to be produced both in bulk and in solution. For the solvent effects in blends, three groups of solvents for blending mixtures of i-PMMA and s-PMMA can be classified: complexing, weakly complexing, and noncomplexing types. Furthermore, the maximum effect of the mixing ratio on the stereocomplex formation is reported to be for mixtures with a weight ratio of (i/s) = 1/2, although other ratios of (i/s) = 2/1 and 1/1 are also reported to induce complexation. Various explanations have been proposed for the formation of stereocomplexes between i-PMMA and s-PMMA. In 1989, a PMMA stereocomplex model was proposed by Schomaker and Challa that the double-stranded helix structure of PMMA stereocomplex was composed of a 9_1 i-PMMA helix surrounded by a 18_1 s-PMMA helix on the basis of x-ray analysis on the stretched fibers. More recently, Kumaki and co-workers proposed that a triple-stranded-helix model for PMMA complex structure on the basis of high-resolution AFM analysis.

O'Reilly and Mosher were the first to report the conformational energy difference of stereoregular PMMA by using IR spectroscopy. The absorbance band at 860 cm^{-1} corresponding to an s-PMMA vibration is stronger in the spectra of the stereocomplex solution in acetonitrile (a complexing solvent) than in the corresponding solution of only s-PMMA, which corresponds to long s-sequences in the extended chain conformation of the backbone. Earlier, Tretinnikov with Ohta reported that a strong preference of trans–trans backbone conformation both for i-PMMA and s-PMMA; the preference is only slightly higher for sPMMA. Namely the complexation significantly increases the regularity of torsion angles for the backbone trans state in i-PMMA and s-cis → s-trans isomerization of ester groups in s-PMMA.

More recently, the stereocomplex capacity of tactic PMMAs (i-PMMA and s-PMMA) of high molecular weights has been shown to be enhanced by adding a third polymer, which decreases Tg of the originally rigid PMMA chain segments.

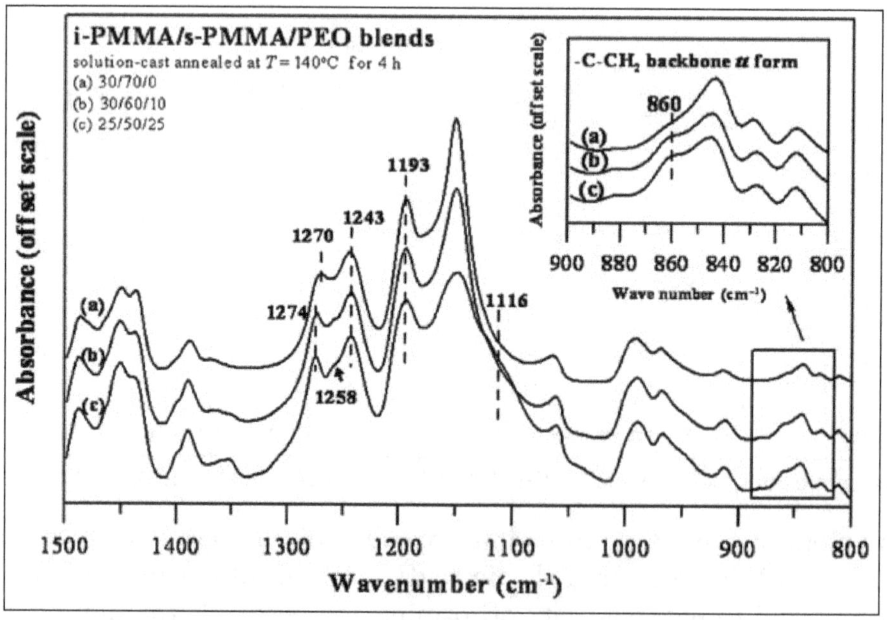

FT-IR spectra in 1000–1500 cm^{-1} for solution-cast ternary i-PMMA/sPMMA/PEO blends annealed at 140 °C for 4 h.

Chang and Woo have reported that a physical interaction or miscibilityenhanced chain entanglement probably exists between PEO and the stereocomplex of tactic PMMAs, leading to stereocomplex formation in the i-PMMA/ s-PMMA/PEO ternary blend. Figure shows FT-IR spectra in 1000–1500 cm^{-1} for solution-cast ternary i-PMMA/s-PMMA/PEO blends annealed at 140 °C for 4 h. The characteristic bands of complexes formation at 1258 and 860 cm^{-1} are clearly observed in a ternary i-PMMA/s-PMMA/PEO blend but are absent in the i-PMMA/s-PMMA blend, indicating that the complexes do form in greater extents via PEO in the blend.

Stereocomplexes of i-PMMA or s-PMMA with diblock- or triblockcopolymer have been reported, such as polyisobutylene block copolymers, poly(dimethylsiloxane) block copolymers, and polybutadiene (PB) block copolymers. Yu and Jerome have reported that stereocomplexation of tactic polymers increases the tensile strength of PMMA–PB–PMMA copolymers, containing small s-PMMA blocks at constant molecular weight of PB and iPMMA, whereas an opposite effect is observed for longer s-PMMA blocks.

Tactic Polypropylenes (a-PP, i-PP, s-PP)

Polypropylenes, depending on synthesis routes, can exist in three different tacticities: atactic-, isotactic-, and syndiotactic-PP (a-PP, i-PP, and s-PP). These three tactic forms of PP have very different physical and mechanical properties. a-PP, of M_w = 12,000, is noncrystalline and amorphous, with T_g = ca −27.2 °C (measured with DSC) and density = 0.85 g/cc. i-PP has T_g = −26 °C and T_m = 160 °C (measured with DSC), and density = 0.9 g/cc. i-PP is known to pack into two crystal cells = α and β forms. The α-form crystal in i-PP is monoclinic with 3/1 helix conformation, whereas the β-form crystal in i-PP is hexagonal cell with same 3/1 helix conformation. Syndiotactic polypropylene (s-PP), at M_w = 127,000 and M_n = 54,000 g/mol, has T_g = −4.3 °C, T_m = 126 °C, and density = 0.9 g/cc. s-PP usually has higher T_g than i-PP, but s-PP has a lower Tm than i-PP. Of these three forms, only i-PP is commercially important as it has a higher melting point than s-PP. Commercially produced PP usually exhibits a mix tacticity composed of mostly isotactic segments with a small percentage of syndiotactic isomers. Their structures with dangling bonds on two ends are illustrated below.

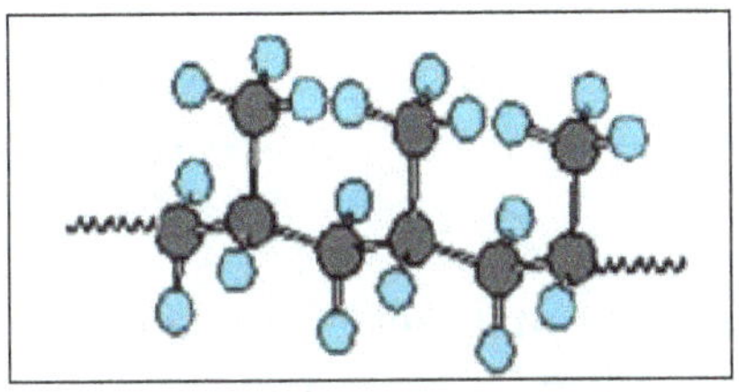

Isotactic polyropylene (i-PP).

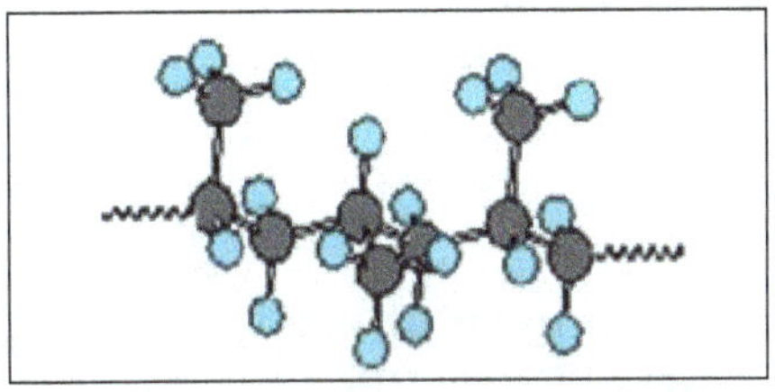

Syndiotactic polypropylene (s-PP).

Blends of polymers of same chemical structures but different configurations may have a good likelihood of forming homogeneous mixtures among the stereoisomers, or occasionally even crystalline complexes, but they are not always miscible. Crystalline complex formation in mixtures of i-PP/s-PP has not been reported; the blend miscibility and phase behavior between PPs of different tacticities, however, have been widely studied. Although blends of polypropylenes of different tacticities have been widely studied in the past, there are still on-going debate on the true phase behavior. That is, what is the phase behavior for the blends of PP of different tacticities. It is critical and interesting to probe whether the PP blends show miscibility or immiscibility with UCST. Except for several earlier theoretical predictions based on the Flory–Huggins mean-field theories, UCST behavior had not been experimentally proven for blends of s-PP/i-PP or a-PP/s-PP, owing to interference of the correct interpretation on the phase behavior from PP crystallinity.

Semicrystalline polymers with different tacticities (i.e, stereoisomers) are not all miscible, which seems a peculiar phenomenon considering that tactic polymers possess identical chemical units differing merely in configurations. For a long time, it has been an intellectually interesting quest that intrigues many investigators probing for answers. A lengthy debate is ongoing whether or not two semicrystalline polymers of same chemical structure but different tacticity are truly miscible in the amorphous state, since they are of the same chemical structure. Further, the problem is compounded by the fact that it is usually difficult to assess precisely the phase behavior or miscibility in mixtures of polymers differing only in tacticity but with closely spaced, or almost the same glass transition temperatures. The phase behavior in melt-state binary blends of tactic polypropylenes (i-PP, s-PP, and a-PP) has long been a difficult subject because all constituents possess same chemical units with same physical properties other than different crystals at solid state. Note that the blend of two isomers, such as PVAc/PMA, differs from the blend system of two tactic polymers, such as i-PP and s-PP, in that the former is polar and amorphous and T_gs of two constituents are far apart, but the latter is nonpolar and semicrystalline with closely-spaced T_gs, and thus more difficult to interpret the phase behavior. Earlier, Thomann and co-workers have claimed that blends of i-PP/sPP are immiscible. In addition, Maier and co-workers have concluded that polypropylenes (PPs) of different tacticities (a, i, or s) are not all miscible with each other. They have concluded that the a-PP/i-PP blend system is miscible, but the a-PP/s-PP system is incompatible with separate phases. Silvestri and Sgarzi have concluded that a-PP is partially miscible with s-PP but immiscible with i-PP, which contradicts with those found by Maier and co-workers or Thomann and co-workers.

A study on morphology of tactic polypropylene mixtures and reported that the i-PP/a-PP blend has UCST below 155 °C and is miscible at a melt state, whereas i-PP/sPP mixtures exhibit a phase-separated texture at crystallization temperatures. The result of Philips agrees with that of Maier and co-workers, but directly contradicts that of Silvestri and Sgarzi. These findings illustrate unsettled issues on blends of PP. The phase behavior in the i-PP/s-PP blends may be difficult to deal with, owing to extremely fast crystallization of both constituent polymers (i-PP and s-PP). Interpretation of the miscibility or phase behavior at the melt state of the tactic PP mixtures based on morphology of the annealed crystals at crystallization temperature may be misleading, and conflicting views have been presented.

Woo and co-workers have utilized novel approaches to construct experimental UCST phase diagrams of blends of tactic PPs by separating the amorphous phase domains from the crystalline spherulites, yielding plausible data for experimentally determining the UCST in s-PP/i-PP blend versus a-PP/s-PP blend, rather than theoretical predictions of phase diagrams. Figure shows the experimental UCST for a s-PP/i-PP blend upon heating to high temperatures above their melting points. Theoretically computed phase diagrams for blends of tactic PPs are also shown in the figure for comparison with the experimentally defined phase diagrams. In addition, the study of Woo and co-workers used the method of equilibrium melting points to estimate the Flory–Huggins interaction parameter for the a-PP/i-PP blend and it has been shown to possess a significantly negative value (χ_{12} = −0.21). This value of interaction strength proves that the a-PP/i-PP blend is indeed miscible in the melted amorphous as well as semicrystalline states, However, the interaction parameters for the s-PP/i-PP and a-PP/s-PP blends have been found to be nearly zero (χ_{12} = −0.02 and −0.007, respectively, at T = 150–180 °C), indicating that the interactions in the two latter blends are rather weak and that the corresponding phase behavior for the s-PP/i-PP and a-PP/s-PP blends may border on immiscibility at ambient temperature.

In comparison with blends of nonpolar PP, blends of polystyrenes of three different tacticities (a-PS, i-PS, s-PS) have also been widely studied. The phase behavior and miscibility in blends of PSs of different tacticities have been shown to differ from those in blends of PPs with different tacticities, perhaps owing to the fact the former ones are more polar in molecular structures but the latter ones are nonpolar possessing only weak van der Waals forces.

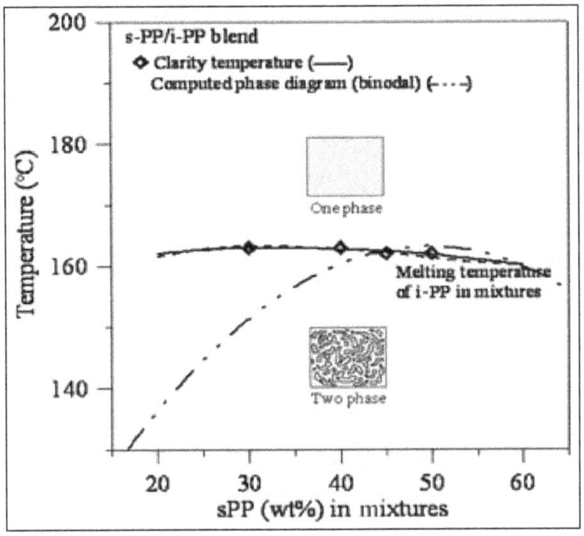

Experimental UCST diagrams for s-PP/i-PP blends in comparison with theoretical UCST prediction. Crystal melting points (for higher melting i-PP) plotted in the graph form.

In addition, the crystallinity in PP blends might have made it more difficult to interpret the miscibility and phase behavior in the amorphous domains. By contrast, the i-PS/s-PS blends, with much lower crystallization rates and higher T_gs than those of PP blends, can be quenched to a fully amorphous state and preserved at ambient for convenient characterization of the phase behavior of the blends at amorphous state. Yeh and co-workers proposed miscibility in the a-PS/i-PS blend system based on results of a crystallization kinetic study; they, however, did not provide more direct evidence other than an analysis of the crystallization kinetics of the blend. s-PS and its blends have been an intensive focus of various studies. Ermer and co-workers have tentatively proposed that s-PS and a-PS might be at least partially miscible based on the results of solvent diffusion behavior in the blends. A study has followed up some points and positively proven miscibility in the a-PS/s-PS blend by utilizing examination of the interaction parameter from measurements of the equilibrium melting point from the thermodynamics point of view. Interactions between the isomeric polymers, even being miscible, are expectedly low.

Earlier, Runt investigated a classically known miscible blend system of a-PS/i-PS (M_w ca 50,000 for both) and found that the polymer– polymer interaction parameter (χ) for the a-PS/i-PS pair is about −0.003, which essentially is zero, indicating very weak interactions between the components. An earlier study has shown that the interaction parameter in the a-PS/s-PS system is similarly a small negative but slightly a greater value ($\chi = \sim -0.1$) than that for the a-PS/i-PS system studied by Runt. The phase behavior of blends of different tactic polymers is influenced by the structural parameters such as tacticity, chain polarity, and size of pendant groups. It is of interest to compare the phase behavior and miscibility in various blend systems of polymers differing only in types of tacticities. For the case of polystyrenes with a relatively bulky pendant group and more polar by nature, the miscibility in blends of tactic polymers has been proven to be independent of tacticity in polystyrene. On the other hand, for polypropylenes with a less bulky CH_3 pendant group (i.e, lower T_g and faster crystallizing), and less polarity and higher crystallinity (than polystyrenes), the miscibility and phase homogeneity in binary tactic PP blends are more influenced by tacticity in PP. In general, there seems no single rule for predicting the phase behavior in blends composed of polymers of same chemical structures differing only in tacticity, especially when more phenomenal factors are present to influence the phase behavior and intermolecular interactions.

Tactic Polystyrenes (a-PS, i-PS, s-PS)

In contrast to i-PP being the most widely used commercial plastic among the three tactic forms of PP, a-PS is the most useful one among the three tactic configuration, owing to its high enough T_g (~100 °C) of a-PS even though it is fully amorphous and does not have any crystallinity. In fact, a-PS is one of the most widely used plastics for almost 50 years since its first commercialization. Two tactic forms of PS, semicrystalline i-PS and s-PS, were latecomers in comparison to atactic PS. Unlike i-PP being a useful commercial product among the three tactic forms of polypropylenes, the isotactic form of PS (i-PS), however, has little commercial applications, owing to the much lower T_g and not high enough crystallinity in i-PS in comparison to s-PS. Not until its successful commercial synthesis via stereospecific polymerization in as early as 1985, s-PS was increasingly studied. A structural difference of tacticity leads to variation in crystal packing. i-PS is packed into a trigonal crystal cell with 3/1 helical conformation, but s-PS is known to packed into several crystal forms (or called modifications) depending on thermal or solvent treatments.

Unique to s-PS is its polymorphism in crystal cell types under different thermal histories. Most other semicrystalline polymers possess only one type of unit crystal cell; for others, polymorphisms with two crystal cells are known to exist. s-PS exhibits complex polymorphism and multiple melting endotherms in association with crystal cells or lamellar patterns. Generally speaking, four crystal forms (α, β, γ, and δ) in thermal- or solvent-treated s-PS are known. A combination of α" and β' crystals is produced when melt-crystallized at low temperatures and the fraction of β' -type crystal increases with successive increases of isothermal temperatures. Only β (β')-type crystal is generated in s-PS if melt-crystallized at high temperatures above 260 °C with no α crystal (neither α' nor α"). Melt-crystallization of s-PS is known to pack into a mix of two common crystal cells: α and β-forms. The α-form s-PS is hexagonal and is favored kinetically, whereas the more stable β-form s-PS is an orthorhombic cell with a zig-zag planar conformation. The relative fraction of α"- and β' -type crystals can be influenced as a function of isothermal melt-crystallization temperatures. In addition, the maximum melting temperatures erasing thermal history and the residence time in the melt also have a significant effect on the polymorphism of s-PS. Furthermore, the polymorphism state of s-PS can be altered in the presence of other miscible polymers: for example, poly(2,6-dimethyl-p-phenylene oxide; PPO) or atactic polystyrene (a-PS). Both the α"- and β' -type crystals are observed in the neat melt-crystallized s-PS; however, only the β' -crystal is found in miscible s-PS/a-PS or s-PS/PPO when crystallized at the same conditions. Apparently, blending with the miscible amorphous polymer reduces the possibilities of the formation of α-type crystal in s-PS. In addition, cold-crystallization (in contrast with melt-crystallization) has a dramatically different effect on the polymorphism of s-PS.

Only the α-type (containing two modifications α' and α") crystal with no β-type crystal generates if s-PS cold-crystallized from its amorphous glassy state. Many studies investigated phase transformation in s-PS by using FT-IR and/or Raman spectroscopy, DSC, and x-ray analysis, and so on. Correlations between the polymorphism and multiple melting endotherms have been an interesting subject of studies. Coexisting α"- and β' -type crystals (or other additional crystal forms) inevitably lead to multiple melting peaks upon scanning; moreover, the presence of polymorphic crystals in s-PS can further complicate the melting and thermal behavior. Interpretation of the phase transition (transformation between different crystals) may be additionally complicated by a phenomenon of multiple-melting behavior in s-PS.

Synthesis of tactic polystyrene and its structure/property have been two main focuses of s-PS. Synthesis of s-PS requires specific catalysis. Four crystal types (α, β, γ, δ) and some mesophases/submodifications have been identified in s-PS. The α and β forms are the main crystal packing in thermally processed s-PS, whereas the γ and δ crystals are identified only in the solvent-treated s-PS. The δ and γ forms in the solvent-treated s-PS are of a monoclinic crystal cell (with helical chain conformation), and the cell dimensions depend on the types and amount of residual solvent that may be trapped in the crystal. The δ and γ-crystals in the solvent-treated sPS are more like mesophases that transform readily to the α, β", or β' crystals upon heating the solvent-treated s-PS to high temperatures near melting. The melt-crystallized α and β crystals in s-PS are the main subject related to the thermal behavior, crystal structures, thermodynamics, kinetics, and stability of these two major crystal packing (α vs β) in s-PS.

Analyses of melting behavior, diffractograms, or IR spectra, and so on of s-PS can be complicated by the presence of coexisting polymorphic crystals. In general, a total of four melting peaks (labeled as P-I, P-II, P-III, and P-IV from low to high temperatures) have been identified in a melt-crystallized s-PS that typically contains mixed fractions of both crystals. By refining the techniques of obtaining s-PS with individually isolated α- or β-crystals, recent studies have been able to correct suspected inaccuracy of some thermodynamic and kinetic measurements in earlier studies and to interpret the relative stability of the various crystals in s-PS. s-PS samples could be prepared so that they contained purely isolated α or β crystal, and the individual crystal types are used for more precise characterization for analysis. The P-I and P–III melting peaks are attributed the β' crystal, whereas P-II and P-IV are attributed to the α" type.

In addition, kinetic and thermodynamic characterizations have been thoroughly performed on individually isolated crystal types and reveal that the α crystal of s-PS has a lower melting temperature than the β crystal, with T_m, α° = 281.7 °C, but T_{m,β° = 288.7 °C. The crystallization kinetics of the α-crystal is a heterogeneous nucleation with higher rates, whereas the β crystal is a homogeneous nucleation with lower rates. The βtype is more thermodynamically stable than the α" type. But the α" type is kinetically more favorable. Both α to β crystals are stable solids, and transformation between them can only be achieved by melting and repacking, but not by solid–solid transformation. Nevertheless, the individual melting/reorganization of these two crystals might undergo crystal transformation via the solid–liquid–solid transition. The crystallization kinetics of β' crystal is a homogeneous nucleation and lower rates. By comparison, crystallization kinetics of the α crystal is a heterogeneous nucleation and higher rates. Microscopy characterization also revealed a highly nucleated crystallization of the α crystal. The effect of blend miscibility on the polymorphism behavior in s-PS is also discussed. Effects of miscibility on polymorphism in s-PS have been investigated by studying two miscible blends of s-PS/a-PS or s-PS/PPO. Both miscible blends containing s-PS have been found to favor growth of β crystal than neat s-PS when subjected to the same melt-crystallization conditions.

The γ and δ crystals are less encountered and usually involved only in solvent treatments of s-PS; melt-crystallization of s-PS usually leads to the α-type (a hexagonal or trigonal unit cell) and β type (an orthorhombic cell) commonly seen in bulk s-PS subjected to thermal treatments (annealing, cooling from molten state, isothermal cold- or melt-crystallization). The α crystal in s-PS has a lower melting temperature (T_{m,α° = 281.7 °C) than the β crystal ($T_{m,,\beta^\circ}$ = 288.7 °C). The heats of fusion measured at the respective maximum crystallinity are also different between these two crystals: $\Delta H_{f,\alpha}$ = 27.9 J/g for the α crystal, which is lower than $\Delta H_{f,\beta}$ = 31.5 J/g for the β crystal, at the respective maximum crystallinity. The results suggest that the β' type is more thermodynamically stable than the α" type; therefore, when crystallized at higher temperatures closer to the equilibrium melting, β' crystal is the preferential species. In addition, molecular weights in polymers may affect the polymorphism in s-PS. Using s-PS of a low molecular weight (M_w = 63,000), it has been found to develop only a β-type unit cell upon melt crystallization at all temperatures. A correlation between the lamellar morphology and multiple-melting behavior in the s-PS has been demonstrated by using the sole β-form crystalline s-PS (Mw = 63,000). When crystallized at lower temperatures (250 °C or lower), the β-form s-PS displays three multiple melting peaks (ie labeled as P-I, -II, -III) and the spherulitic morphology is composed of two main crystalline domains, which are initially thin and straight lamellar bundles (P-1) and eye-like lamellae platelet crystal (P-II). The P-III crystal was not initially present in significant extent in s-PS crystallized at low temperatures. However, upon scanning to higher temperatures, the initially thinner flat-on lamellae (P-I) are melted and then immediately repacked to more thickened and branched lamellae (P-III), whose orientation is also twisted from the original flat-on to an edge-on (perpendicular to substrate) pattern. Detailed mechanisms of transformation between P-I and P-III crystals via melting/repacking are evident.

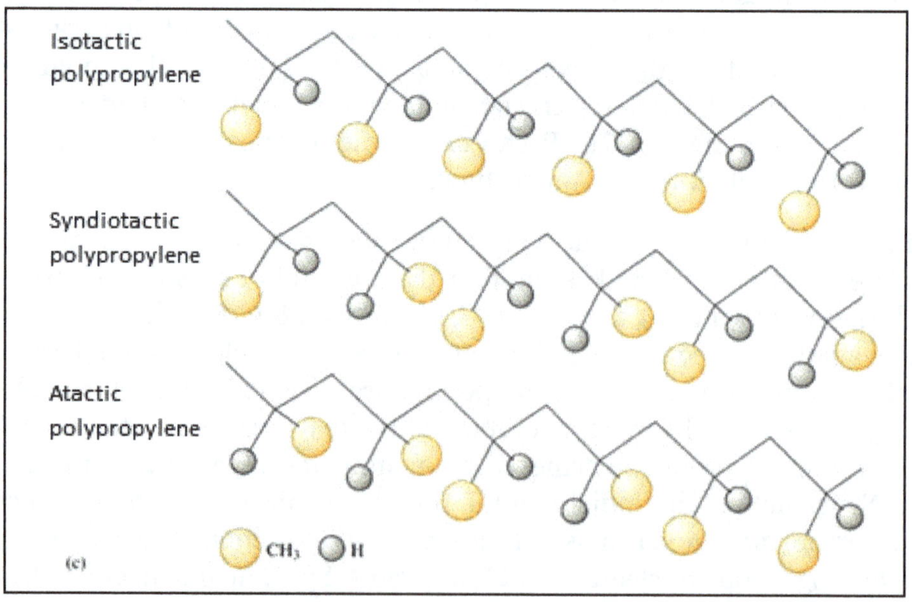

P-III lamellae could also be directly developed in sPS by melt-crystallization at higher temperatures (260 °C), whereas P-I and P-II crystals were absent. Only the P-III crystal (branched, thick, axialite) exists in the s-PS melt-crystallized at 260 °C or above. In addition, polymorphism in s-PS may be influenced by cold- or melt-crystallization, which, respectively, is the crystallization from a glassy and melt state of the polymer. Melt-crystallization in s-PS usually produces mixed fractions of both α- and β-form crystals, except for low-Mw s-PS that melt-crystallizes into predominantly β-form crystal. A study by Sun and Woo has pointed out that cold-crystallization of s-PS produces only the α-type unit cell packed into crystalline domains of different spherulitic morphology patterns. The morphology contains a granular-spherulitic texture when cold-crystallized at low temperatures, whereas high-temperature cold-crystallization produces additional sheaf-like lamella radiating out from the central spheres.

References

- Stereochemistry-of-polymers, events-list: polymerchemistry.euroscicon.com, Retrieved 24 March , 2019
- Introduction-polymers, chemistry, science-maths-technology: open.edu, Retrieved 1 June, 2019
- Isomerism, polymer%20resources, llever: uscupstate.edu, Retrieved 23 June, 2019
- Polymer-structure, materials-science: sciencedirect.com Retrieved 21 March , 2019
- Cis-Trans, polymer%20physics: polymerdatabase.com, Retrieved 12 April, 2019

Permissions

All chapters in this book are published with permission under the Creative Commons Attribution Share Alike License or equivalent. Every chapter published in this book has been scrutinized by our experts. Their significance has been extensively debated. The topics covered herein carry significant information for a comprehensive understanding. They may even be implemented as practical applications or may be referred to as a beginning point for further studies.

We would like to thank the editorial team for lending their expertise to make the book truly unique. They have played a crucial role in the development of this book. Without their invaluable contributions this book wouldn't have been possible. They have made vital efforts to compile up to date information on the varied aspects of this subject to make this book a valuable addition to the collection of many professionals and students.

This book was conceptualized with the vision of imparting up-to-date and integrated information in this field. To ensure the same, a matchless editorial board was set up. Every individual on the board went through rigorous rounds of assessment to prove their worth. After which they invested a large part of their time researching and compiling the most relevant data for our readers.

The editorial board has been involved in producing this book since its inception. They have spent rigorous hours researching and exploring the diverse topics which have resulted in the successful publishing of this book. They have passed on their knowledge of decades through this book. To expedite this challenging task, the publisher supported the team at every step. A small team of assistant editors was also appointed to further simplify the editing procedure and attain best results for the readers.

Apart from the editorial board, the designing team has also invested a significant amount of their time in understanding the subject and creating the most relevant covers. They scrutinized every image to scout for the most suitable representation of the subject and create an appropriate cover for the book.

The publishing team has been an ardent support to the editorial, designing and production team. Their endless efforts to recruit the best for this project, has resulted in the accomplishment of this book. They are a veteran in the field of academics and their pool of knowledge is as vast as their experience in printing. Their expertise and guidance has proved useful at every step. Their uncompromising quality standards have made this book an exceptional effort. Their encouragement from time to time has been an inspiration for everyone.

The publisher and the editorial board hope that this book will prove to be a valuable piece of knowledge for students, practitioners and scholars across the globe.

Index

www.ingramcontent.com/pod-product-compliance
Lightning Source LLC
LaVergne TN
LVHW080648210926
848485LV00003B/344

* 9 7 8 1 6 4 7 2 8 0 1 5 4 *